Warning:
Psychiatry Can Be
Hazardous to Your
Mental Health

Also by William Glasser, M.D.

Reality Therapy: A New Approach to Psychiatry (1965)

Schools Without Failure (1969)

Positive Addiction (1976)

The Quality School: Managing Students Without Coercion (1990)

Choice Theory: A New Psychology of Personal Freedom (1998)

Getting Together and Staying Together (2000)

Counseling with Choice Theory: The New Reality Therapy (2001)

Unhappy Teenagers: A Way for Parents and Teachers to Reach Them (2002)

Warning: Psychiatry Can Be Hazardous to Your Mental Health

William Glasser, M.D.

Foreword by Terry Lynch, M.D.

WITHDRAWN

📖 HarperCollins*Publishers*

HarperCollins books may be purchased for educational, business, or sales promotional use. For information, please write: Special Markets Department, HarperCollins Publishers Inc., 10 East 53rd Street, New York, NY 10022.

FIRST EDITION

Designed by Nancy Singer Olaguera

Printed on acid-free paper

Library of Congress Cataloging-in-Publication Data

Glasser, William.
    Warning: psychiatry can be hazardous to your mental health / William Glasser; foreword by Terry Lynch.—1st ed.
        p. cm.
    Includes bibliographical references.
    ISBN 0-06-053865-1
    1. Psychiatry. 2. Mental illness—Alternative treatment.
    I. Title.

    RC465.5.G536 2003
    616.89—dc21

                                                    2002038735

03 04 05 06 07 ❖/RRD 10 9 8 7 6 5 4 3 2 1

To Brian Lennon, my esteemed colleague from Ireland.

Since we met many years ago, I have felt your mind is a mirror of mine. When you sent me Terry Lynch's book, *Beyond Prozac: Healing Mental Suffering Without Drugs,* I was on my way.

Acknowledgments

First of all to my wife, Carleen. She was with me every step of the way. If you want to know what I think of her, read the dedication to my book *Choice Theory*.

To Al Siebert of Chapter 13, for his courage to stand by his convictions, even though psychiatrists at the Menninger Clinic sent him to a mental hospital for articulating his strong belief in the concepts of choice theory well before I articulated them.

To Terry Lynch, M.D., for writing a powerful foreword and supporting the value of choice theory, even though he had never been exposed to it before.

To Peter Breggin, M.D., for his willingness to take on both the psychiatric drug manufacturers as well as the psychiatric establishment and lay much of the framework for this book. It seemed a good omen when I found out we shared an outstanding medical school, Case Western Reserve, and the same birthday.

To Jon Carlson, for all the support he has given me since we met. He helped me become an honorary Adlerian. To the American Counseling Association (ACA), for their willingness to honor me three years running as a major speaker and give me a platform for choice theory.

To Monika Kahlenbach, Dan Joynt, Linda Catrabone, and others, all members of the ACA from Connecticut who are working hard to establish the ideas of this book in their state.

To Linda Harshman and the staff of the William Glasser Institute.

To a group of thirteen actors who, playing the roles of unhappy people, created a Choice Theory Focus Group, which we videotaped. These sensitive actors were able to bring the choice theory concepts to life in front of your eyes.

Finally, to the guys I play tennis with who are catching on to what I do. Several years ago they told me they had seen the movie *Babe* and encouraged me to see it by saying, "There is a pig in that movie that seems to practice what you preach."

Contents

Foreword

by Terry Lynch, M.D.

The message conveyed by Dr. William Glasser in the title of this book, that psychiatry can be hazardous to your mental health, may seem shocking. You may reasonably ask, how could it be that a specialty we widely accept as being authoritative on mental health is actually the opposite—hazardous to the people it treats? But Dr. Glasser is a psychiatrist with over forty years' experience in the field. As the founder of reality therapy, accepted internationally as a valid and effective form of therapy, he is not to be dismissed lightly.

Dr. Glasser is not alone. A small but significant number of doctors and others working in health care fields are very concerned about the direction in which psychiatry is heading. I am one such medical doctor, having worked as a family physician in Ireland for almost twenty years. When I qualified for my M.D. degree, like my colleagues I was a fervent believer in the medical approach to mental health. I continued to be so for about ten years after qualifying. Then I began to have doubts. Over a seven-year period I researched the beliefs and practice of psychiatry. The more I researched, the more concerned I became. I discovered that medical pronouncements regarding the scientific basis of psychiatry are far more dubious than the public believes. I realized that my medical training in the area of mental health had been hugely deficient.

I could not continue to work within the narrow approach

to mental illness that prevailed in the medical community and live with my conscience, so, in 1999, I retrained in psychotherapy, and I began to build my working arrangements around what my patients needed rather than what we doctors believe is best for them. My questioning culminated in the publication of my book, *Beyond Prozac: Healing Mental Health Suffering Without Drugs,* in 2001.

The medical approach to mental distress is based on unproven hypotheses, in particular the theory that the fundamental cause of mental distress is biological, either a biochemical imbalance, a genetic defect, or both. Psychiatry has convinced itself and the general public that this hypothesis is not a hypothesis but a proven fact. In doing so, modern psychiatry has made a major error of judgment, an error so fundamental that it should never occur in any discipline purporting to be scientific. But psychiatry gets away with it, because instead of policing psychiatry to ensure that it does not lose the run of itself, legislators and the general public alike place great faith and trust in the integrity and objectivity of psychiatry and psychiatric research. What limited policing there is of psychiatry is not in-depth and relies heavily on the bona fides of psychiatrists and medical researchers.

Decades of intensive psychiatric research have failed to establish a biological cause for any psychiatric condition. The lack of biological evidence is confirmed by the extraordinary fact that not a single psychiatric diagnosis can be confirmed by a biochemical, radiological, or other laboratory test. I know of no other medical specialty where vast numbers of people are treated on the presumption of a biochemical abnormality. The medical profession's reliance on biology as the determining factor for psychiatric disorders is founded upon faith rather than true scientific research. It seems to me that doctors shoot in the dark far more than the public realizes.

And despite all you have heard, little real progress has been made in the research on the genetics of mental illness. As the *Chicago Tribune* stated on March 20, 2001: "It is a fact that despite decades of research, not a single gene responsible for mental illness has been found—the quest has been shattered by the debunking of highly visible reports localizing genes for schizophrenia. Similar fates met discoveries of genes for manic-depression, alcoholism, homosexuality."[1]

There is little enthusiasm within the field of psychiatry for ideas that run counter to conventional medical theory. In the year after *Beyond Prozac* was published in Ireland, in April 2001, I received over two thousand calls and letters from people affirming that the book made great sense to them. This in a country with a population of just over three million. Yet for the most part my best-selling book was greeted with silence by the medical profession. While the book received wide coverage from the general media, no Irish medical newspaper or medical journal reviewed it. Three months after an Irish psychiatrist wrote in an Irish national newspaper that my ideas would not contribute to furthering our understanding of depression, my book was placed on the short list for the United Kingdom's MIND Book of the Year award, the first Irish book to make this list in the twenty-one-year history of this prestigious award.

We doctors have become preoccupied with diagnosing mental illness and initiating medication treatments. The supposed mental illness becomes the focus of attention and the underlying human issues go unnoticed and unresolved. The biochemical model of the illness remains an unproven hypothesis. Since any hypothesis may ultimately be found wanting, I believe that as a society our view of mental health must

[1]Peter Gorner, "Gene Study Doubted Before Ink Is Dry: Search for a Link to Mental Illness Eluding Science," *Chicago Tribune*, 20 March 2001.

expand, as Dr. Glasser suggests, and go beyond the narrow medical approach. There is too much at stake.

It is difficult to find a single aspect of psychiatric practice that is based solidly on science. While doctors vehemently emphasize that the primary solution to the worldwide suicide crisis is the early detection and treatment of depression with medication, research does not support the view that antidepressants prevent suicide.

In 2002 the Royal College of Psychiatrists had to downgrade their pronouncements regarding the effectiveness of antidepressant drugs. For decades, and in particular since the newer SSRI (selective serotonin reuptake inhibitors) antidepressant drugs such as Prozac, Paxil, and Efexor came on the market over ten years ago, psychiatrists have stated with great authority that these drugs were at least 70 percent effective in treating depression. Now the Royal College of Psychiatrists has had to concede that these drugs are roughly only 50 to 60 percent effective. Bearing in mind that the effectiveness of placebos in treating depression is in the region of 47–50 percent, the case for the widespread prescribing of antidepressant drugs is dubious, to say the least. An increasing number of studies are failing to find significant differences between the effectiveness of placebo and antidepressants.

In a startling commentary on modern psychiatric care, schizophrenia outcomes in the United States and other developed countries are far worse than in the poorer countries of the world. In studies of schizophrenia outcomes, the World Health Organization has repeatedly found that recovery from schizophrenia was a far more frequent occurrence in poorer countries such as India, Nigeria, and Colombia than in developed countries like the United States, England, and Denmark. These studies repeatedly found that approximately two-thirds had good outcomes in poorer countries as compared with approximately one-third in developed countries.

Psychiatry has been slow to learn from its mistakes. The medical profession, legislators, and the public alike seem to have forgotten that most of the medications used in the treatment of mental distress over the past one hundred years have been either dangerous, addictive, or ineffective. In the early days it was drugs such as alcohol and opium, later followed by the barbiturates and the amphetamine group of drugs. These were followed in the 1960s by the benzodiazepine group of drugs, which have been in use for the past forty years. Each of these drugs was introduced with great fanfare, and then used for decades as the latest wonder drug for mental distress. It took doctors fifty years to grasp the true addictive potential of the barbiturate group of drugs. Today doctors would balk at the idea of prescribing barbiturates or amphetamines, but in 1967, 23.3 million prescriptions for amphetamines were written in the United States; in that year 12 million people took amphetamines on medical advice.

The medical experts did not want to contemplate the possibility that these drugs might be addictive. Clearly, the medical experts got this very wrong, and there is worrisome evidence that precisely the same scenario is currently unfolding with antidepressant drugs. For years patients have been telling their doctors that they find it hard to come off antidepressants. Currently there is widespread prescribing of antidepressant drugs, drugs that doctors have assured the public are definitely not addictive. Yet the newer antidepressants have not been systematically tested for their addictive potential, surely a gross oversight given the addictiveness of many of their predecessors. What short memories we doctors have when it suits us.

Early in 2002, in what may turn out to be an enormously important development, the manufacturers of the antidepressant Paxil were forced by the U.S. Food and Drug Administration to issue a new warning to patients and doctors, acknowledging that some people get hooked and might suffer severe side effects when they stop taking Paxil. Paxil's manufacturers were found to be in breach of the industry's code of practice by misleading the public. In effect, this ruling means that doctors will now have to acknowledge what they have been vehemently denying for years, that antidepressants are indeed addictive.

The lessons of the past have not been learned because the field of psychiatry does not want to learn them. It is because learning the lessons would mean that psychiatry would have to look at itself, its practices, and beliefs. Psychiatry seems to have forgotten that the primary reason for its existence is supposed to be to serve the public interest rather than its own. Psychiatry has walked itself into a cul-de-sac from which it is unwilling to return. If we don't learn from our mistakes, we keep repeating them.

But there is hope. Throughout the world there is a small but significant group of doctors who are speaking out. William Glasser is one such doctor. In this book, Dr. Glasser illustrates why we ought to be concerned about psychiatry's dominance of mental health care. Refreshingly without medical jargon, he presents the case for reform of mental health care. But William Glasser goes further. He presents us with an alternative approach to mental health care.

William Glasser maintains that since relationships are central to human happiness, recovery should involve exploring how we relate to others and looking for ways of improving how we relate to others, particularly within our closest relationships. You might say that this is just common sense, and indeed it is. Unfortunately in this world, common sense can be an uncommon commodity. The medical profession stoutly refuses to focus on the issue of relationships and how we might relate in more productive ways. Why? Because relating does not fit into the medical model. And anything that does not fit into the medical model is discarded as unworthy of serious consideration. Because, like many other institutions, the institution of psychiatry puts its own survival and advancement above all other considerations.

William Glasser's years of inquiry and questioning have culminated in the development of what he calls choice theory. He cogently argues that what medicine has labeled as mental illness is in reality varying degrees and expressions of unhappiness. He points out that medicine's preoccupation with mental ill-health means that people searching for the opposite—mental health and well-being—receive little meaningful guidance from the medical profession. William Glasser offers this book as a worthwhile guide to mental health.

Dr. Glasser builds his approach on a solid foundation of kindness, respect, support and compassion, human qualities that I have found to be crucially important foundation stones in working with people. Central to this book is Dr. Glasser's belief that a great deal of human unhappiness—much of which subsequently becomes labeled as mental illness by the medical profession—relates to the choices we make. He illustrates that by identifying and owning the choices we make, we become responsible for ourselves. Responsible, that is, in the positive sense of being able to respond and make healthier choices.

I have long been concerned about the medical tendency to disconnect mind and emotions from the body. Doctors treat people's conditions as if they are dealing with a machine, as if no other aspect of the person's being or experience could possibly have anything to do with the condition for which they are being treated. Seduced by the attraction of science and misinterpreting the true meaning of science, a holistic approach to health has been sidelined by the medical profession. This is highly unscientific because it refuses to consider what is obvious. Mind and body do not exist as separate entities; they are intimately intertwined. Yet medicine shows little willingness to explore possible links between mind (psyche) and body (soma).

It is time to examine the psychiatric monopoly on mental health care. Given that more than one in four people come in contact with the mental health services during their lifetime, it is essential that no stone is left unturned in our efforts to create an effective mental health care service. Not only is psychiatry leaving many stones unturned; it is doing its level best to ensure that those stones remain unturned.

For far too long, psychiatry has been allowed to regulate itself. I believe that an independent public inquiry into mental health care is required. Many studies have found that counseling and psychotherapy could make a major contribution in mental health care. But because therapy does not fit into the biological medical model, psychiatry has demonstrated little interest in expanding the role of psychotherapy in treating mental distress. Many medical schools, such as the prestigious Johns Hopkins, no longer require their psychiatric trainees to study psychotherapy. The question therefore arises: How well equipped are doctors to deal with the emotional, psychological, and social problems of their patients?

For decades doctors have dismissed research that challenges the medical model as being flawed. Ironically, over the past few decades it has become clear that psychiatric research has itself been woefully unscientific and seriously flawed. Even in prestigious journals such as the *British Journal of Psychiatry* an estimated 40 percent of research papers contain significant statistical errors. And that's just the statistical errors.

Although the older drugs for the treatment of schizophrenia have been in widespread use around the world since the 1950s, prescribed for tens of millions of people, there seems to be little interest within the medical profession in carrying out a comprehensive evaluation of those drugs. A comprehensive review of them certainly is warranted. The drugs are responsible for the greatest medically induced epidemic of all time, an irreversible and very distressing neurological disorder known as tardive dyskinesia. Tens of millions of people have been treated long-term with these drugs over the past fifty years. Given that 250 out of every 1,000 people taking them for five years can be expected to develop tardive dyskinesia, millions of people worldwide have developed this disorder as a consequence of their medication. These figures are well established and not in dispute.

You might assume that surely doctors would not put their patients at such risks unless the benefits of the drugs were definite and well established. Unfortunately that is not the case. The quality of medical research into these major tranquilizers from the late 1930s to the late 1990s has not quite been up to scratch. A recent major review of two thousand research trials conducted during this period into the treatment of schizophrenia has raised very serious questions regarding the quality and reliability of these studies and these treatments. This major review also included the newer drugs being used to treat schizophrenia.

I am very concerned that unless there is an inquiry into psychiatric practice, the older major tranquilizers will quietly slip into history, to be replaced by the newer drugs. I fear that history will continue repeating itself, that in thirty or forty years' time these newer drugs will then be old drugs that, like their predecessors, are found to have serious side effects, and will be replaced by still newer "wonder" drugs. The wheel goes round and round, unless the authorities say "*Stop!* What's going on here? Are there better, safer ways?"

Like many other groups and institutions, the medical profession holds steadfastly to its belief system. There is much at stake here for the members of the medical profession, who, along with those in the pharmaceutical industry have heavily invested their time, their hopes, their identity, and their lives, into this belief system.

The medical approach is typically to suppress: to suppress anxiety, anger, rage, fear, and other unpleasant emotions people may have. But suppression of symptoms is not sufficient. The suppression may give temporary relief—relief that is often welcome and appropriate—but frequently there is little long-term healing. The person's suppressed emotions and issues are not dealt with, and may be pushed further underground by the drugs.

There is ample evidence that psycho-socio-emotional-relationship-life issues play an important part in the creation of so-called mental illness. We need to focus on helping people get their lives back on track, at a pace that works for them. I believe that choice theory, the subject of this book, has great promise in this regard.

The quality of mental health care, which directly affects 25 percent of the population and affects many more indirectly, is not a fringe issue—it is a major, mainstream issue in society. An independent review of mental health and mental health care is urgently required. Any such review must be wide-ranging, asking fundamental questions about the medical approach to mental illness, the medical belief system. The entire diagnostic process, as well as approaches to treatment, need to be reviewed in detail. Our perception and treatment of so-called mental breakdown needs a comprehensive exploration and review. The views of those at the receiving end of mental health care must be listened to.

I am not rejecting science. I am all for science—true science, that is. To lose sight of the bigger picture, as I believe medicine has in the case of mental health, is also to lose sight of the true meaning of science. Because true science excludes no possibility in its search for understanding and progress.

Preface

The following article from the op-ed page of the March 10, 2002, *New York Times* focuses on schizophrenia. But in this book I go beyond schizophrenia. I explain that none of the people described in the *DSM-IV*,[1] the official diagnostic and statistical manual of mental illness published by the American Psychiatric Association, are mentally ill. I don't deny the reality of their symptoms; I deny that these symptoms, whatever they may be, are an untreatable component of an incurable brain malfunction. I do not see their symptoms as mental illness but as an indication that they are not nearly as mentally healthy as they could learn to be.

What I believe and will explain in depth throughout this book is that the basic human problem has nothing to do with the structure or physiology of our brain. We are by our nature social creatures and to be mentally healthy or happy, we need to get along well with the people in our lives. Unhappy people like John Nash, and all those whose disorders appear in the

[1]The *DSM-IV* is the fourth and latest edition of a large book produced by the American Psychiatric Association in which all the known psychological symptoms are described. In it, these symptoms are grouped together into syndromes, each of which is referred to as a mental disorder. The symptoms described are accurate. Grouping them together and calling them mental disorders is wrong.

DSM-IV, are not mentally healthy. They are lonely or disconnected from the people they need.

As I will explain in depth, they express their unhappiness in the symptoms described in the *DSM-IV* plus many others like pain and anger. For them to become more mentally healthy and happier, we have to offer compassion, social support, education, and counseling. They do not need brain drugs and electric shocks, all of which harm their brains.

If you have difficulty accepting this explanation, check out Chapter 13 and the many references listed in the Appendix. Or to save time, read *Beyond Prozac: Healing Mental Health Suffering Without Drugs,* the seminal work on the hazards of biological psychiatry by Terry Lynch, a man you've already met in the foreword to this book.

Beautiful Minds Can Be Reclaimed

by Courtenay M. Harding

The film *A Beautiful Mind,* about the Nobel Prize–winning mathematician John F. Nash Jr., portrays his recovery from schizophrenia as hard-won, awe-inspiring and unusual. What most Americans and even many psychiatrists do not realize is that many people with schizophrenia, perhaps more than half, do significantly improve or recover. That is, they can function socially, work, relate well to others and live in the larger community. Many can be symptom-free without medication.

They improve without fanfare and frequently without much help from the mental health system. Many recover because of sheer persistence at fighting to get better, combined with family or community support. Though some shake off the illness in two to five years, others improve much more slowly. Yet people have recovered even after 30 or 40 years with schizophrenia. The question is, why haven't we set up systems of care that encourage many more people with schizophrenia to reclaim their lives?

We have known what to do and how to do it since the mid-1950s. George Brooks, clinical director of a Vermont hospital, was using thorazine,[2] then a new drug, to treat patients formerly dismissed as hopeless. He found that for many, the medication was not enough to allow them to leave the hospital. Collaborating with patients, he developed a comprehensive and flexible program of psychosocial rehabilitation. The hospital staff helped patients develop social and work skills, cope with daily living and regain confidence. After a few months in this program, many of the patients who hadn't responded to medication alone were well enough to go back to their communities. The hospital also built a community system to help patients after they were discharged.

These results were lasting. In the 1980s, when the patients who had been through this program in the '50s were contacted for a University of Vermont study, 62 percent to 68 percent were found to be significantly improved from their original condition or to have completely recovered. The most amazing finding was that 45 percent of all those in Dr. Brooks's program no longer had signs or symptoms of any mental illness three decades later.

Today, most of the 2.5 million Americans with schizophrenia do not get the kind of care that worked so well in Vermont. Instead, they are treated in community mental health centers

---

[2] I, Terry Lynch, Peter Breggin, and others listed in the Appendix, believe that the people diagnosed as schizophrenic would have done better without the Thorazine or any other strong psychiatric drug. It would be more accurate to say that they recovered despite the drugs rather than because of them.

that provide medication, which works to reduce painful symptoms in about 60 percent of cases and little else. There is rarely enough money for truly effective rehabilitation programs that help people manage their lives.

Unfortunately, psychiatrists and others who care for the mentally ill are often trained from textbooks written at the turn of the last century, the most notable by two European doctors: Emil Kraepelin in Germany and Eugen Bleuler in Switzerland. These books state flatly that improvement and recovery are not to be expected.

Kraepelin worked in back wards that simply warehoused patients, including some in the final stages of syphilis who were wrongly diagnosed with schizophrenia. Bleuler, initially more optimistic, revised his prognoses downward after studying only hospitalized patients, samples of convenience, rather than including patients who were ultimately discharged.

The American Psychiatric Association's newest *Diagnostic and Statistical Manual, DSM-IV,* published in 1994, repeats this old pessimism. Reinforcing this gloomy view are the crowded day rooms and shelters and large public mental-health caseloads.

Also working against effective treatment are destructive social forces like prejudice, discrimination and poverty, as well as overzealous cost containment in public and private insurance coverage. Public dialogue is mostly about ensuring that people take their medication, with little said about providing ways to return to productive lives. We promote a self-fulfilling prophecy of a downward course and then throw up our hands and blame the ill person, or the illness itself, as not remediable.

In addition to the Vermont study, nine other contemporary research studies from across the world have all found that over decades, the number of those improving and even recovering from schizophrenia gets larger and larger. These long-term, in-depth studies followed people for decades, whether or not they remained in treatment, and found that 46 percent to 68 percent showed significant improvement or had recovered. Earlier research had been short-term and had looked only at patients in treatment.

Although there are many pathways to recovery, several factors stand out. They include a home, a job, friends and integration in the community. They also include hope, relearned optimism and self-sufficiency.

Treatment based on the hope of recovery has had periodic support. In 1961 a report of the American Medical Association, the American Psychiatric Association, the American Academy of Neurology and the Justice Department said, "The fallacies of total insanity, hopelessness and incurability should be attacked and the prospects of recovery and improvement though modern concepts of treatment and rehabilitation emphasized."

In 1984, the National Institute of Mental Health recommended community support programs that try to bolster patients' sense of personal dignity and encourage self-determination, peer support and the involvement of families and communities. Now there are renewed calls for recovery-oriented treatment. They should be heeded. We need major shifts in actual practice.

Can all patients make the improvement of a John Nash? No. Schizophrenia is not one disease with one cause and one treatment. But we, as a society, should recognize a moral imperative to listen to what science has told us since 1955 and what patients told us long before. Many mentally ill people have the capacity to lead productive lives in full citizenship. We should have the courage to provide that opportunity for them.

> *Courtenay M. Harding is a senior director of the Center for Psychiatric Rehabilitation at Boston University's Sargent College of Health and Rehabilitation Sciences. Copyright © 2002 The New York Times Company. Reproduced here with permission.*

This book is my attempt to provide this opportunity for the majority of people diagnosed with disorders listed in the *DSM-IV* and to provide it free or at very small cost.

Warning:
Psychiatry Can Be
Hazardous to Your
Mental Health

one

Who Am I, Who Are You, and What Is Mental Health?

In the forty-five years I've been in psychiatric practice, I have worked in every area of psychiatry except with small children (when consulted about a small child, I work with the parent or parents). During this time I've become more and more convinced that both adolescents and adults with psychological problems can be taught, through the way I counsel, to improve their own mental health and become much happier than they were.

But also during this time, I've observed that the idea of mental health, never a strong component of psychiatry, has disappeared altogether. What the vast majority of my profession, which in this book I will call the *psychiatric establishment*, does today is diagnose people displaying symptoms as mentally ill and prescribe psychiatric drugs to treat them. These psychiatrists call themselves biological psychiatrists and some, who use brain drugs exclusively, call themselves psychopharmacologists.

If you have any psychiatric symptom, such as those

described in detail in the *DSM-IV*, there is no longer any concerted effort from this psychiatric establishment to establish a doctor-patient relationship and counsel you about what's on your mind. You are told that your mental illness is caused by an imbalance in your brain chemistry that can only be corrected with drugs. This practice has grown to the point where I believe the title of this book is understated. The few psychiatrists who still counsel almost always combine this effort with psychiatric drugs and many believe the drugs are the most important component of their treatment.

What the present psychiatric establishment has done that can harm your mental health extends far beyond the psychiatrist's office. Now almost all health professionals are caught in this neurochemical "web."[1] Brain drugs dominate the entire "mental health" landscape.

To give you an example of the magnitude of this domination, in the year 2001, 111 million prescriptions were written for just one class of drugs, selective serotonin reuptake inhibitors or SSRIs such as Paxil, Prozac, and Zoloft. This represents a 14 percent increase over the year 2000 and the percentage is still growing (as reported in the Health section of the July 1, 2002, *Los Angeles Times*). Recent studies show that this class of drugs may be no more effective for depression than placebos.[2]

General practitioners, as much as or more than psychiatrists, are diagnosing mental illnesses and prescribing Prozac and other similar brain drugs. Pediatricians are diagnosing

---

[1]For details of this "web," read "Prescription for Scandal" by Anthony Black, which is reproduced in Chapter 13.

[2]Irving Kirsch, Thomas Moore, Alan Scoboria, and Sarar Nicholls, "The Emperor's New Drugs," *Prevention and Treatment*, Volume 5. Article 23. Published on Web site (Journals.APA.org), July 15, 2002.

attention deficit disorder (ADD) or attention deficit hyperactive disorder (ADHD) in your children and prescribing Ritalin, a strong synthetic cocaine that acts on your child's brain in ways that are not yet known and may never be known. Psychologists, social workers, and counselors are diagnosing mental illnesses and teaming with general medical practitioners as well as psychiatrists to get brain drug prescriptions for their clients. Often this is done without the prescribing doctor examining the people they prescribe for in any depth.

These drugs are not harmless. There is a large body of scientifically sound psychiatric research that lays out in detail the harm these drugs can do both to your mental health and to your brain itself. At the same time, this research points out that these drugs are nowhere nearly as effective as is claimed by the companies that make them. There is a dark side to biological psychiatry you may never have heard about. You will hear about it in this book.

Still, it might be argued that it is worthwhile risking the damage these drugs may do to your brain if there are no safe, effective alternatives to them. But there are. Quick, effective counseling without brain drugs has advanced beyond what it was twenty-five years ago. The problem is that most of the people who need counseling can't afford what it costs to talk to a counselor, much less a psychiatrist. Their health insurance will cover brain drugs for years on end but rarely more than a few counseling sessions.

Damaging as this practice may be, the real horror of this system is the harm it does to our innate desire to try to take care of ourselves. The message that has now come through loud and clear in the media is that *when you are diagnosed with a mental illness there is nothing you can do to help yourself.* The message of this book is that no matter what mental illness you or a family member may be diagnosed with, there is a lot you can do to help yourself or a member of your family who needs help.

The media went "gaga" when John Forbes Nash Jr. recovered from schizophrenia, a supposedly incurable mental illness that, even with the best psychiatric care, separates its sufferers permanently from reality. But as you read in the article from the *New York Times* that I cited in the preface, this is not the case at all. Many psychiatrists,[3] like myself, don't believe schizophrenia is a mental illness. It is one of the thousands of ways that unhappy people like Nash deal with their unhappiness.

No psychiatrist did much for John Nash. What he did to recover, with the help of his wife and the tolerance of the Princeton math department that let him wander its halls for years, he eventually did for himself. Unfortunately, near the end of the movie about his life a blatant untruth was introduced when it was stated that his unanticipated recovery was greatly furthered by the use of modern brain drugs.

What is written in his biography,[4] and shown somewhat in the movie, *A Beautiful Mind,* is that he did not take his brain drugs regularly before 1970 and after that year took none at all. I think it is more accurate to say his much later recovery was aided by the happiness of being awarded the 1994 Nobel Prize for economics and the fact that his wife did not give up on him. His recovery occurred despite his psychiatric care, not because of it.

As important as counseling is, and I have spent my whole career counseling, the thrust of this book goes much further. In it I will examine the concept of *mental health* in depth and suggest that, just as physical health can be taught to millions

---

[3]One of the first psychiatrists to deny the existence of mental illness was Thomas Szasz. His warning that this is a mistake was published in *The Myth of Mental Illness,* (New York: Paul Hoeber, 1961).

[4]Sylvia Nasar, *A Beautiful Mind,* (New York: Simon and Schuster, 1998).

of people who are out of shape, but not physically ill, mental health can be taught to millions of people who are unhappy but not mentally ill.

I began to think this was possible as soon as I started my counseling practice[5] in the late 1950s. I became involved teaching counselors who worked with delinquent teenagers to help them become much more law-abiding than they had been. No one called them mentally ill but it was obvious they were struggling with unhappiness. Soon after that, I began to teach schoolteachers to use the same ideas to help the many unhappy students they faced daily in their classrooms. Both groups found the ideas helped them to succeed with young people they had not succeeded with previously.

Seeing how effective these ideas were, they began to use them at home. Over and over they both told me and wrote to me about how much better they were getting along with their wives, husbands, and children. Everyone was happier. By the early 1960s I put these ideas together and in 1965 I published the book I am best known for, *Reality Therapy*.

Reality therapy has grown to the point where it is now taught all over the world by the hundreds of instructors who teach for the William Glasser Institute.[6] We work successfully in every aspect of mental health and education and we do it without recommending that anyone be given brain drugs. In my practice I have never prescribed a brain drug no matter how severe the symptoms of the psychological problem.

[5]In this book I use the terms *counseling, therapy,* and *psychotherapy* interchangeably. I prefer the term *counseling* because that term is less associated with brain drugs.

[6]The work of this Institute is described in the Appendix. It has its own Web site (www.wglasser.com).

Almost from the date *Reality Therapy* was published, I began to get supportive feedback from professionals who read it. They told me essentially what their colleagues had been telling me since the 1950s. They were getting help themselves from what they were using to counsel others. It took some years for me to realize that this helpful, but unforeseen, effect could be expanded and clarified into a new psychology, I call *Choice Theory*®.

As soon as I began to teach choice theory, the feedback to me and to my instructors about the value of applying it to their personal lives increased. After I wrote the basic book, *Choice Theory,*[7] in 1998, even though I had not written it as a self-help book, I began to hear from many more people, professionals and nonprofessionals, about how valuable choice theory was in their own lives and with members of their family.

Their letters and e-mails were filled with a variety of comments that stressed how happy they were since they'd begun to use choice theory to guide their lives. As this continued, I realized that choice theory could be the road to mental health that I had been searching for since I began the study of psychiatry.

I now believe in the following metaphor: *Happiness is mental health.* And also in its opposite: *Unhappiness can never be mental health.* This doesn't mean that mentally healthy people are happy all the time. But when they are unhappy they can learn to use choice theory to help themselves and often succeed. Although I will explain happiness in detail later, I will define it briefly here so you can take a look at your life and see how close you are to this definition.

[7]This is the book that got me started thinking about mental health: William Glasser, *Choice Theory,* (New York: HarperCollins, 1998). See Chapter 14 for more details of this book as well as the eight others I have written since 1998.

- *Happiness or mental health is enjoying the life you are choosing to live, getting along well with the people near and dear to you, doing something with your life you believe is worthwhile, and not doing anything to deprive anyone else of the same chance for happiness you have.*

Since I am writing this book for you to use on your own to help yourself or an unhappy family member or friend, I want to tell you who I think you are[8] so you can see if you are reasonably close to the profile of the people I am writing it for.

I believe you are a person who is looking for a way to find more happiness in your life without depending on prescribed brain drugs such as Prozac or self-prescribed ones like alcohol. You are not averse to going to a counselor but on your tight budget your eyes are always open to ways to find happiness by your own efforts. I see you as a thoughtful person who enjoys finding out more about yourself and how to use this understanding to get along better with the people near and dear to you.

I also see you as a person who is willing to try something new such as choice theory as long as you understand what you are doing and that you can stop any time you want. Further, I see you as a caring person who recognizes when someone dear to you is unhappy and needs help. This book will explain how to help this person and improve your own mental health in the process.

Later in the book, as I explain choice theory, I will conduct several Choice Theory Focus Groups for people who have read this book and want to get together with others to discuss how they can incorporate choice theory into their lives and become happier. In these groups they can learn to help themselves and

---

[8]In this book I will not attempt to use both gender pronouns whenever there is no clear choice, but sometimes I will use *he*, sometimes *she*. This will avoid the clutter of obsessive political correctness.

others. The group may need someone who knows choice theory to get them started, but after a few meetings the group should be able to continue on its own. As you read about these focus groups, you will see that they are not therapy groups. They are called Choice Theory Focus Groups because the participants focus on learning to use choice theory to improve their mental health.

Focus groups have no set beginning or ending and participation is completely voluntary. They can go on as long as each person in the group wants to attend. I see people coming and going to these groups just as people come and go to AA meetings. I suggest that these groups stay small, maybe ten to fifteen members, but that's up to each group. Any reader might want to join a group or start one. In order for you to get a feel for the kind of people in a group you'll meet later in the book, I'd like to introduce you now to its members.

Bev, a forty-four-year-old single mother who is depressed over her inability to get along with her totally out-of-control seventeen-year-old daughter, Brandi.

Jill, a forty-year-old family physician who suffers from migraine headaches. She is aware there is no physical cause for these headaches.

Molly, a thirty-two-year-old married woman who suffers from fibromyalgia. This diagnosis means all of her muscles hurt, along with other discomfort but, like Jill, no physical cause can be found for her symptoms. Molly doesn't believe her doctor. She believes there is a physical cause for her symptoms that has yet to be discovered.

Amy, a twenty-eight-year-old single woman who suffers from panic attacks but doesn't want to take drugs or get professional help. She's been trying to figure out what to do for herself on her own.

Neil, Amy's thirty-four-year-old brother who has read some of the early chapters of the book and persuaded Amy to join the group to see if she could get some help.

Jeff, a thirty-year-old man who has suffered from rheumatoid arthritis since he was a teenager and has heard that improving his mental health might help his arthritis.

Barry, an angry, controlling thirty-eight-year-old man who is unhappy in his marriage.

Joan, Barry's thirty-four-year-old wife who is also unhappy in her marriage.

Selma, A fifty-four-year-old divorced woman who is the mother of Jim, a thirty-one-year-old man who was diagnosed with schizophrenia when he was nineteen and who lives at home. She joined the group to see if she could find out what more she could do for him.

Roger, Joan's father, a sixty-one-year-old, happily married man and family counselor who's been using my ideas for years and was invited by Joan to join the group. Roger likes the ideas of mental health and is willing to help in any way he can.

Professionals, like Roger, are welcome in the group but they must accept the rule that they can't charge for anything they contribute. If, however, the group gets to know and trust the professional and a member wants to consult with him or her privately, the group member should realize that this would not be a free consultation.

What I will try to teach in this book is that there is a vast difference between seeing yourself as mentally ill, believing you can't help yourself, and seeing yourself as unhappy, but tending to believe you can help yourself. As unhappy as you may be, if you can see yourself as helping yourself or someone else, this book is for you.

If, however, you are presently on psychiatric drugs or being counseled for what you have been told is a mental illness, you may have two questions. *Can this book help me? And would I be welcome in the focus group you just described?* My answer to both these questions is a resounding *yes*. Learning to use what I teach in this book is not an either or proposition. You can learn it and put it to work in your life no matter what you've been told and whether or not you are taking brain drugs or being counseled. I believe any focus group you'd like to join will be glad to have you. But you should understand from the start: *These are not counseling groups.*

The people in them are not there to hear extensively about your past or present unhappiness, for example, how much trouble you are having getting along with a spouse, parent, child, or boss. Or about how much pain you are suffering or how unfair life has been to you. They will be interested in hearing about how you are applying the choice theory ideas of this book to your present problems. And in helping you learn to do this more effectively as the group continues to meet.

I've also had to assure some counselors who have expressed concern over whether the ideas in this book will make it harder for them to make a living. Basically they've said, if people can use these ideas to help themselves, what will happen to us? I am flattered and encouraged by their concern. But, it isn't realistic. Counselors would be overwhelmed with demands for their services if even a small percentage of all unhappy people could be persuaded to give up ineffective psychiatric drugs. According to a 1999 report to the Surgeon General by a blue-ribbon committee of establishment psychiatrists, 28 percent of the adults in the United States carry *DSM-IV* psychiatric or addictive diagnoses. In the year 1999, we're talking about 75 million unhappy people. Add to this number, children, parents like Bev and Selma, as well as people with chronic pain such as Jill and Molly and this number could easily be much higher.

Add all the unhappy people in prison,[9] all the unhappy children who don't do well in school, many of whom are labeled ADD and ADHD, and all the unhappy married people like Barry and Joan, who don't come close to warranting a psychiatric diagnosis, and you can see that there are easily 100 million people who need help. If the opportunity to join Choice Theory Focus Groups never materializes, almost all these unhappy people will do as they are doing now: trying to live with their unhappiness or settle for the placebo effect of brain drugs with all their side effects. Any unhappy person who reads this book should be able to help himself or a person close to him to better mental health at, essentially, no cost to anyone .

This book is my attempt to *disconnect* mental health from mental illness treated with brain drugs and make mental health more available to everyone. It is all about learning to use choice theory in your life. There is no counseling or any other treatment offered. You are not risking anything by trying what I suggest. From years of experience, I can state without reservation that, while I can't guarantee better mental health, I can guarantee that nothing in this book can harm you.

[9]Right now essentially no effort is being made to improve the mental health of anyone in prison. Without this book, we have taught choice theory to prisoners and they have shown great interest. Follow-up from the Oklahoma prison where it was taught seems favorable. It can do no harm and might do a great deal of good. The cost would be negligible. See the appendix for details.

two

The Difference between Physical Health and Mental Health

29 September 2001
Victoria, Australia

Dear Dr. Glasser,
*   It was with great interest that I listened to your interview on television while you were on a visit to Australia. We have in our family, severe psychological problems with depression being addressed with drugs of various kinds and it hurts us deeply to watch this intelligent person (my brother-in-law) being ravaged by their effects. He seems to know about choices in life but is so deeply entrenched in finding justifications for these "quick-fix cures" that giving himself choices is the hardest decision he can make.*

*   What can we do? I've read your book* Choice Theory *but we need help and I am hoping that you can point us in the right direction—possibly you have clinics either in New South*

*Wales or, preferably, Victoria, where he can attend or you may
know someone here that practices this theory.
I look forward, with interest, to your reply.*
Yours truly,
Judy S.[1]

I get letters like this from all over the world from people who
have become involved with establishment psychiatry. Judy's concern after hearing me speak on Australian television was obvious.
Her brother-in-law was being treated with brain drugs that were
harming him. The same thing that is harming him is happening
to people all over the world. In this chapter I will begin to explain
what can be done for him directly to improve his mental health
and, if nothing can be done directly, how she can help him.

Because most people know a lot about physical health but
very little about mental health, I will start this chapter by
explaining the difference between them. Comparing them and
seeing how they differ will help you to understand more about
both. I think you'll be surprised by this difference.

The Physical Health Continuum

When health is discussed in the media, the focus is almost all on
curing and preventing physical illness as if our population is
either physically ill or physically healthy. This focus is extremely
misleading because the vast majority of us are neither physically
ill nor physically healthy. Only a very small percentage of our
population is so physically ill that they have been given a medical
diagnosis such as cancer, heart disease, or diabetes. Even a

---

[1] I sent her the name of a counselor near where she lives. When this
book is finished, I'll send her a copy.

smaller percentage of our population is so physically healthy they are fit to run a marathon. This leaves millions of us who are neither physically ill nor in top physical condition.

Therefore, if we want to describe the physical health of our whole population, the best way to do this is to use a continuum with relatively few of us at either end and most of us somewhere in the middle. This continuum is shown below:

Physically Ill-----Out of Shape-----Physically Healthy

Physical illness based on pathology is shown at the left; physical health at the right. But the best way to describe the vast majority of us who occupy the middle is being out of shape. Still, in the affluent, indolent, well-nourished society we live in, I think it's fair to say that most of us in the middle would like to be more physically fit. And almost all of us are well aware of what we need to do to get there: exercise more and eat less.

But from experience, we are also painfully aware that knowing what to do is far easier than doing it. To get encouragement, attention, and instruction, we may enroll in a fitness class or employ a personal trainer. But whether we do it alone or with help, we still have to do it. No one or no medication can do it for us.

With that in mind, I'd like to move to the point of this book by offering a mental health continuum analogous to the physical health continuum.

Mentally Ill-----Unhappiness-----Mentally Healthy

On the left are a relatively few mentally ill people who suffer from medically recognized brain pathology. Their illnesses correspond to physical illnesses such as cancer, heart disease, and diabetes. Examples are Alzheimer's disease, Parkinson's disease, epilepsy, brain tumors, or multiple sclerosis. Any neurology text will list these and many more. Pathology occurring in your brain has a lot to do with your being unhappy, but these diseases are not how we express unhappiness. They are diagnosed and treated by neurologists, not psychiatrists.

The "mental illnesses" that establishment psychiatrists diagnose, treat, and list in the *DSM-IV* should not be labeled illnesses, because none of them is associated with any brain pathology. See the Appendix for confirmation of this claim. They are the many ways unhappy people express their unhappiness, which I will start to explain in this chapter and continue throughout the book. As you can see in the continuums shown above, the mental equivalent of out of shape is unhappiness. Here in the middle of each continuum, whether you are out of shape or unhappy, if you know what to do and are willing to do it, you can help yourself move toward the health end of the continuum and you won't need medication to do it.

The difference is if you are out of shape, you know what to do: exercise and lose weight. If you want information on how to do it there are hundreds of reliable books you can read that teach you what to do. If you are worried that you are too out of shape to exercise, you can go to your doctor for a physical examination. If he can't find any pathology, he'll tell you you're not sick, and to just take it easy for a while and check back with him if you're uncomfortable.

But if you have symptoms of unhappiness such as anxiety or depression and you want to improve your mental health to get rid of them, very few of you know what to do. You may look for a book, but as far as I have found, there are few books that specifically teach you how to improve your mental health. Right now, if you want to move toward the right end of this continuum, you are very much on your own.

If you are expressing your unhappiness with symptoms such as depression or any other of the symptoms described in the *DSM-IV,* you are in the middle of the mental health continuum and are not mentally ill. Here, you are much worse off than if you are in the middle of the physical health continuum. You not only don't know what to do, but if you seek help from an establishment psychiatrist, you may end up worse off than before you sought help.

Instead of assuring you that you are not mentally ill, as your medical doctor will do if you're out of shape, he will tell you that you are mentally ill, that there is pathology in your brain or in your brain chemistry. He will explain that to cure the "pathology" you do not have, you need brain drugs. He may also tell you that you may also benefit from counseling, but he will stress that the brain drugs are the important part of the treatment.

It will not occur to him that your brain is normal or that what's wrong is you're unhappy, and your symptoms are your way of expressing your unhappiness. Instead he will try to convince you and your family that you are on the left side of the continuum: you are mentally ill. He sees everyone with symptoms such as depression, or any others listed in the *DSM-IV,* as mentally ill and in need of medication.

By putting drugs into your brain that interfere with its normal functioning, he is a hazard to your mental health. By down-playing what counselors can do for you, he is a further hazard to your mental health. But by far the greatest hazard that he poses to your mental health is his finding mental illnesses that do not exist and, in doing so, robbing you of a chance to do something for yourself.

I have explained the mental health continuum so you can begin to understand what is going on. This book is my attempt to teach you that there is a lot you can do for yourself when you are unhappy or to help an unhappy member of your family at no risk to anyone. Just as there are many books that will teach you how to move toward the top of the physical health continuum, this book can teach you how to do the same for your mental health.

When your establishment psychiatrist makes his diagnosis, especially if your symptom is anxiety or depression, he will tell you that a neurochemical imbalance that exists in your brain is causing your symptoms. The fact that he hasn't a shred of valid evidence to support his claim doesn't bother him. His common sense tells him it is impossible for you to have symptoms such as those described in the *DSM-IV* and still have a physically and chemically normal brain.

In making the case for mental illness, the psychiatric establishment has replaced science with common sense. If you have symptoms, something must be wrong with your brain. Since no reputable scientist has ever found anything pathological in your brain structure, biological psychiatrists focus on what is fleeting, rapidly changeable, and can't be seen under a microscope: abnormal brain chemistry. Since your brain chemistry must change continually as your behavior changes, you can't have the same brain chemistry when you are happy as you have when you are fearful, angry, or depressed. But because it changes does not make it abnormal.

To prove what they claim is true about your brain chemistry, they employ pseudoscience and say your brain chemistry is congruent with your brain activity. They then scan your brain activity and show that parts of your brain are more or less active when you are depressed, fearful, or angry. They then take a huge intuitive leap and claim that the change in brain activity they have just scanned represents your ever-changing brain chemistry. Then they take a further leap and conclude that it was the change in brain chemistry that is causing your fear, anger, or depression.

That conclusion is about as scientific as me taking your heart rate when you are calm and then pointing a gun at you, shooting a few bullets past your ears, taking your heart rate again, and then telling you that you have heart disease because it is now beating abnormally. In this scenario it would only be abnormal if it remained the same.

There is another huge difference between being physically ill and mentally ill that for your own protection or for the protection of a member of your family, you should be aware of. If you have a physical illness like clogged coronary arteries, the doctor offers you a specific diagnosis and an effective surgical treatment. He may also offer you a drug like a statin to lower your cholesterol. But he will not try to force this treatment on you.

When your psychiatrist diagnoses you as mentally ill, especially when the diagnosis is schizophrenia, he is almost certain to tell you that you have pathology in your brain to support his diagnosis and then prescribe drugs to treat it. If you disagree and resist, he may do something no other doctor would do: try to force you to take the medication he believes you need even if he has to have you locked up so you can be watched.[2] To get you locked up he has to declare that you are a danger to yourself or to others in your community. Some people diagnosed with schizophrenia may be dangerous but generally this is not a condition associated with danger.

By far the most dangerous people in any community are unhappy young men between the ages of eighteen and thirty. They are especially dangerous when they drink but no one suggests locking them up or even restricting their access to alcohol. Given the kind of treatment discussed in the Foreword and Preface—kindness, support, compassion, and protection—few people, no matter what their mental illness diagnosis is,

[2]Rather than spend any more time here discussing how your freedom is in danger whenever an establishment psychiatrist diagnoses you with schizophrenia, I have asked my colleague Dr. Al Siebert, a prominent clinical psychologist, to tell you about his personal experience with this diagnosis. See the first part of Chapter 13 for his story.

can be accurately assessed as dangerous enough to have to be forcibly medicated or locked up.

When your psychiatrist tells you or your family that this forceful treatment is necessary, he is not knowingly avoiding the truth. He believes what he's telling you is the truth: that you are mentally ill, there is pathology in your brain, and you need psychiatric drugs, incarceration or both. If you show him the evidence gathered in the Appendix of this book that supports what I claim about mental illness and medication, he will tell you that this research is wrong.

He may add that he has better or more recent brain research to back up what he's telling you. But I advise you to take what he says with a grain of salt. There is overwhelming evidence in the appendix that shows that the research he is citing is funded by the companies that make the drug they are researching. This is about as valid as the research funded for years by the tobacco companies that concluded "scientifically" that cigarettes were neither addicting nor harmful.

If this were just an academic argument about the validity of the diagnosis of mental illness, I wouldn't be writing this chapter. But what's at stake is not academic; it's your mental health, or the mental health of a family member or a good friend. If you take brain drugs, your physical health is at risk, too. I will spend the rest of this chapter, explaining how this unproven psychiatric dogma is hazardous to your mental health in a variety of ways.

Still, as I said in Chapter 1, if you are taking a psychiatric drug, I do not advise you to stop it if you or your family is convinced this drug is helping you. But if you do stop it, do it slowly, because an abrupt withdrawl from these strong, brain-altering drugs may also be harmful to your mental health. But as I said in Chapter 1, where I introduced the hazards of brain drugs, everything I teach about mental health in this book will help you whether you are taking brain drugs or not. But it will work better if you are not on them.

If what I suggest in this book helps you to help yourself or helps you to help a member of your family or a friend, it is important that you trust in your own experience. Don't put what I suggest to work in your life for very long unless you feel better. If you feel better, your mental health is improving. Also, you are not endangering your mental health by doing anything I suggest in this book. None of the people in the Choice Theory Focus Group I cited in Chapter 1, who are starting to help themselves and help each other, is on brain drugs.

Why Psychiatry Maintains the Fiction of Mental Illness

If you read the references in the appendix carefully, you will find out that when you are diagnosed with a mental illness, such as depression, schizophrenia, bipolar disease, or obsessive compulsive disorder, and treated with a brain drug, you have become one of the millions of geese who lay golden eggs for the multibillion-dollar brain drug industry. There are big bucks in brain drugs.

This industry, which masquerades as mental health's best friend, generously funds a variety of groups and activities that promote mental illness and brain drugs. Examples of this funding are lucrative research grants to psychiatrists who can come up with supportive research, plush psychiatric conferences, liberal grants to mental health associations that vigorously support mental illness and brain drugs, large grants to patient advocacy groups that do the same, and millions of dollars to fund high-powered public relations firms to promote the "new drugs" to cure "mental illness" and to persuade the media to report these cures.

In sharp contrast, there is no money from the mental health industry for supporting the mental health paradigm I explain in this book, or for those participating in the Choice Theory Focus Groups I introduced in Chapter 1, who are putting it to work in their lives. The last thing the psychiatric establishment and the drug companies want is for you to get the idea that you can improve your own mental health or help your loved ones to improve theirs at no cost to yourself. You need all the money you can spare to improve your own life, not to improve the bottom line of a huge corporation.

It May Be Easier to Go from Unhappiness to Mental Health Than from Out of Shape to Physically Fit

When you are in the out-of-shape middle section of the physical health continuum, you are not in pain. You enjoy eating and sitting around. It feels good to stay in the middle where you are. You may discover that when you make the effort to diet and exercise, it's difficult and painful. But if you give up, you're ashamed and feel guilty. So as long as we're not sick, most of us are content to live a sedentary life, top off our bellies, and hope for the best.

But if you're in the middle of the mental health continuum, you are unhappy and often have painful symptoms like depression and anxiety. Once you accept that you are not mentally ill and that there are easy-to-use ways to move to the right on the mental health continuum and feel better than you're feeling now, you have a good incentive to keep trying. When you suffer from a mental symptom or are diagnosed with a mental illness, your problem is not that you aren't willing to try to help yourself. You'd like to help yourself. You just don't know what to do, and few people in the mental health field are making any effort to teach you what to do. As I just said, the money is in mental illness, not mental health.

So you have some idea of what you will gain by making an effort to move up the mental health continuum. Let me add to the brief description of mental health that I offered in Chapter 1 and tell you where you'll be if you move all the way to the right on the mental health continuum. Unlike exercise and dieting, each step you take in that direction feels good and increases your incentive to go further.

You are mentally healthy if you enjoy being with most of the people you know, especially with the important people in your life such as family and friends. Generally, you like people and are more than willing to help an unhappy family member, friend, or colleague to feel better. You lead a mostly tension-free life, laugh a lot, and rarely suffer from the aches and pains that so many people accept as an unavoidable part of living. You enjoy life and have no trouble accepting that other people are different from you. The last thing that comes to your mind is to criticize or try to change anyone. You are creative in what you attempt and may enjoy more of your potential than you ever thought was possible. Finally, even in difficult situations when you are unhappy—no one can be happy all the time—you'll know why you are unhappy and you'll attempt to do something about it. You may even be physically handicapped, as is actor Christopher Reeve, and still fit the criteria above.

If that's where you are now, I welcome you as a reader. Not because you need what I suggest, but because you're the kind of person the world needs. For you, helping the people around you to be mentally healthier is an enjoyable part of your life and I hope you find this book useful as you deal with people who need to improve their mental health. Roger, who joined the group I cited in the last chapter and whom you will meet later in the book, is this kind of person. I was glad to have him on board.

What I am trying to do with this book is encourage you to be aggressive in protecting yourself from wrong diagnoses and harmful brain drugs by learning how to be happier and more mentally healthy. I hope you'll share your success with family and friends. But even with what you know now, you should ask any doctor who diagnoses you as mentally ill and advises you to take drugs, to explain what he's basing his diagnosis and treatment on. And are there any alternatives to what he's suggesting that don't involve drugs?

Acquaint yourself with a few of the books in the Appendix so you can see for yourself that there is support for what I suggest. As I said in the Preface, if you have time for only one, I strongly recommend *Beyond Prozac: Healing Mental Health Suffering Without Drugs* by Terry Lynch, M.D., whom you met in the Foreword. I just loved that book. If I ever get seriously unhappy, I'm going to camp on Terry's doorstep.

Be Prepared to Help Yourself

In the last twenty-five years, the psychiatric establishment has almost completely changed direction. It no longer supports the belief, commonly held for centuries and as sensible today as ever, that if you are capable of carrying on a conversation, you should seek counseling when you are unhappy.

In this book, I take this belief a step further. If you are capable of reading and talking about what you've read, there is a good chance you can learn to help yourself by reading this book and talking to others about what you and they can do to use these ideas. As I said in the first chapter, there are millions of unhappy people who will never get counseling but could easily and pleasurably join a cost-free Choice Theory Focus Group.

The current psychiatric belief is that even if you are in good contact with reality and could be counseled, you are still mentally ill and are better treated (in many instances should only be treated) with psychiatric drugs. The establishment psychiatrist you are almost sure to see if you belong to an HMO may be interested in counseling you but is rarely given permission to spend the money to do it. From doing little more than reviewing a brief checklist of your symptoms, he will tell you or your family that you are mentally ill or suffer from a mental disorder.

This now dominant psychiatric practice, financed by a multi-million-dollar media blitz paid for by the drug industry, has been so successful that it has been accepted not only by most psychiatrists but by most medical doctors, as well as many psychologists, social workers, and counselors. It is the way that most HMOs deal with unhappiness and it is embraced by a general public that has no easy access to the truth. The public likes the simplicity of the argument: If you have psychological problems, you are ill; all you need to get your mental health back is a pill.

The public has no awareness that the price of this pill is to blind you to the lessons of this book: You can pursue happiness and mental health on your own. There is a further price you risk when you take strong brain drugs; many of them harm the brain and cause real mental illness.

This harm may be called side effects by your doctor, but once these chemicals are in your brain there is nothing "side" about them. In the book by Joseph Glenmullen,[3] cited in the appendix, he points out many of these side effects and also explains that some of them start when the drug is discontinued so that even getting off the drug is not always safe. The worst side effect he discusses is called tardive dyskinesia.[4] If

---

[3]Joseph Glenmullen, *Prozac Backlash* (New York: Simon and Schuster, 2000).

[4]Tardive dyskinesia is a real brain-damaging mental illness, caused by a variety of brain drugs, and it is very hard to predict who will get it or how much of the drug it is safe to take. The person with it loses control over many of his muscles, including his facial muscles, and writhes and grimaces uncontrollably. In many cases it seems incurable. This is discussed in Glenmullen's book and in books by Peter Breggin, M.D. See the appendix for more details.

this affliction is even a remote possibility, you should pay very close attention to the title of this book.

The chances of getting a physician today, medical or psychiatric, to suggest counseling instead of medication are very slim. The possibility that the doctor will go further, as I do in this book, and suggest there are things you can learn to help yourself toward better mental health without drugs or counseling is virtually nonexistent. He is unaware of the unhappy middle ground you are mired in and from which, with some help, you could learn to extricate yourself.

As much as your doctor or psychiatrist has difficultly accepting the concept of a mental health continuum, he will easily understand the concept of a physical health continuum. Very likely he uses it to improve his own physical health. When he's out of shape, he knows he's not ill and has no need for drugs. He will diet and exercise like anyone else.

The way establishment psychiatrists conclude a person is mentally ill is from the cluster of symptoms the patient complains about or the symptoms his family describes. They get additional information from observing the person's behavior. For example, if the person they see is withdrawn, suspicious, and complains of hearing voices and the family confirms this, he is diagnosed as schizophrenic. Then, when the family asks the psychiatrist where the schizophrenia or other mental illness came from, he tells them it is caused by a brain chemistry imbalance or some other pathology in the brain.

Right now no one knows specifically what causes the symptoms described as illnesses in the *DSM-IV*. There are a lot of inferences such as lowered serotonin, a brain chemical found to be lower than normal in stressed rats who "appear" to the researchers to be depressed. They use rats because the only accurate way to determine serotonin levels in any brain is to grind it up and assay the ground-up material.

But even if the inference in the comparison of rats to humans is correct, no one yet knows whether the depression lowered the serotonin or the lowered serotonin caused the depression. The psychiatrists who use brain scans are guessing at the latter and trying to persuade you to go along with their guess.

In a later chapter on creativity, I'll offer you a plausible explanation for the cause of many of the hard-to-understand psychological symptoms such as hallucinations, delusions, and mania. I believe they are created, and understanding your creativity can help you or your family deal more effectively with this unwanted creativity.

When you are diagnosed as having a mental illness that needs a drug to cure it, the psychiatrist treats you as if you're helpless: there is nothing you can do for yourself. This flies in the face of overwhelming evidence (again see the Foreword, Chapter 14, and the Appendix) that many people, even those with the symptoms called schizophrenia, have helped themselves. They have done so either completely by themselves or with the help and support of friends and family, and their symptoms have disappeared, whether or not they have ever taken drugs. There is also extensive evidence that counseling without drugs can be both effective and longer lasting than if drugs were used instead of or even along with the counseling.

Unlike your psychiatrist, when your medical doctor tells you that your symptoms are caused by an illness such as heart disease, she has explicit pathological proof for her diagnosis from one or more of the following: *physical* pathology observed in her examination, pathology found in *X-ray, CAT scan,* or *MRI* procedures, *microscopic* pathology seen on slides, and/or *chemical* pathology derived from testing your blood or other bodily fluids.

When your psychiatrist tells you that your symptoms are caused by a mental illness, he hasn't a shred of similar evidence. To label a person mentally ill, which now translates in almost everyone's mind as some sort of brain pathology, is to *stigmatize* millions of people who should not be subjected to the rejection, ostracism, harmful drugs, and brutal electric shocks to the brain that can and often do accompany this erroneous label.

Unusual, crazy, or frightening as these symptoms may be, they are no more caused by mental illness than what Timothy McVeigh[5] did was caused by a mental illness. He turned out to be the most dangerous American who ever walked the streets of our country but no one called him mentally ill. What people who are not mentally healthy have in common with each other is their behavior is not predictable. The behavior of happy people is very predictable.

The people who knew him before and after he committed the crime were well aware that McVeigh was unhappy. What they didn't know was what he was planning to do. There is no limit to the destructiveness of unhappy people, just as there is no limit to the kindness, caring, and self-sacrifice of mentally healthy people. In both cases, their brains are normal; it is how they choose to use them that is off the norm.

For example, when your computer fails to do what you want it to do, it is a thousand times more likely that the trouble will be in the way you are using it, or in the software, than in the machine itself. You need to use it more accurately or find better software. You don't need to fix its mechanism any more than you need to "fix" a normal brain with drugs or electric shocks.

A few pages back in this chapter, I offered you a description of a mentally healthy person. For many of you, this may have seemed too good to be true. But if you look around, you will find there are plenty of people among your family or friends, who come close to fitting that description. What they

---

[5]McVeigh blew up the federal building in Oklahoma City on April 19, 1995, killing 169 people. He was executed for this crime.

are doing is what I suggest you can learn to do on your own or with the help of a friend or family member.

Examples of the Physical and Mental Middle Ground

Suppose you are Jon. You're overweight and begin to tire on the golf course by the time you start the second nine. You say to yourself, it's enough. I've got to get into shape. Your friends tell you, before you do anything drastic see your doctor to make sure you're okay. You go to see him, he gives you a complete physical, and tells you the good news: you're out of shape but you're not physically ill. He refers you to a couple of good books on diet and exercise and tells you to start slow and don't overdo it. The only medication he may offer you is vitamins.

You follow his advice and, in a year, you're in better shape than you were when you left high school. It took a little discipline but what to do was clearly explained in the books you consulted. You didn't go out and buy any fancy exercise equipment or expensive health food. You joined a gym, more for the companionship than for the equipment, ate good food from the supermarket, walked the treadmill, and began to do a little running when the weather was good. The path to good physical health was clear and you took it. It wasn't easy but you made the effort and succeeded.

But let's look at a mental health scenario. You are Joan. You've been married eighteen years and your marriage is no longer satisfying. You and your husband argue a lot, too little sex is a problem for him, and too little affection is a problem for you. It's clear to you that your situation at home is beginning to affect your twelve-year-old son's performance in school. The school calls you in and tells you he has stopped paying attention and can't sit still in class. They intimate he needs medication and you should see your pediatrician.

Your HMO pediatrician listens to your story for fifteen minutes and talks briefly to your son who, in essence, tells him he hates school, it's boring, and all his teacher does is lecture and expect him to memorize material he soon forgets. The pediatrician tells you that your son has ADD or maybe even ADHD and prescribes Ritalin.[6] He asks you about your marital situation. When you tell him it's not good and you think it's affecting your son, he suggests you see your doctor and maybe get some medication for yourself.

When you see your medical doctor and tell him you've lost interest in life and feel depressed, it's hardly a new story to him. Since he hasn't the time to inquire about your marriage or your life in general, he focuses on your symptoms. You may even tell him you're unhappy in your marriage. He tells you that it's likely you're suffering from clinical depression, and suggests you take Prozac.

You ask if you need to see a psychiatrist and he says it might be a good idea and refers you to one. It takes three weeks to get an appointment and, when you do, the psychiatrist listens for less than twenty minutes, backs up your doctor's diagnosis, and prescribes Prozac. Neither doctor spent much time inquiring about your lack of love and your problem with sex and, especially, your relationship with your husband.

Your son hates taking the Ritalin but he does do a little better in school; he's more able to sit still. But he seems to have

---

[6]ADD and ADHD stand for attention deficit disorder and attention deficit hyperactive disorder and are commonly treated with a synthetic cocaine derivative called Ritalin. For a description of how harmful this practice is to your child, log on to the Web site of Fred A. Baughman Jr., M.D. (www.adhdfraud.org). Dr. Baughman is a leading critic of the use of brain drugs for all mental illnesses in the *DSM-IV.*

lost his spark, his eyes have a kind of a dead look, and you're worried about that. The Prozac gives you a lift but what little sex drive[7] you had seems to have almost disappeared. You feel better but your marriage is worse. What crosses your mind is that you've put all your eggs and your son's eggs in the doctors' baskets and all you have to show for your visits are two bottles of pills.

When his doctor examined Jon, he understood his situation immediately. Jon was told that he was out of shape but he wasn't sick and didn't need medicine. What he needed to do was get involved in a physical health program. If he were willing to do that, his problem would be solved. Diet and exercise programs made sense both to him and to his doctor. He followed them and his fitness improved markedly. But as I stated earlier, he had to do it. No doctor or drug could do it for him.

When Joan and her son saw the doctors, she was told, without a thorough effort to get to know her or her son, that she and her son could do nothing for themselves. They were suffering from a diagnosable mental disorder caused by a neurochemical imbalance in their brains. Each of them was given brain medication to correct the imbalance.

In both instances, little attention was paid to their complaints: *he hated school* and *she lacked love in her marriage.* Joan was told—she was *told,* it was not merely *suggested*—that what was wrong with her and her son's brains could be diagnosed from the presenting symptoms and treated with drugs to correct the problem.

But suppose Joan had run into a doctor who did not believe that there was anything physically or chemically wrong

---

[7]Drugs such as Prozac, Zoloft, Luvox, or Paxil have the effect of reducing the sex drive in many people. Look at the warning that comes with your prescription.

with her brain, that what she needed was counseling. The kind of counseling I use, reality therapy, which I have developed and taught over the last forty years, is very effective. Using it, I have successfully counseled people who have been diagnosed as Joan and her son were, plus many with much more severe diagnoses such as schizophrenia, without using psychiatric drugs.

As successful as it can be, counseling paid for by your HMO is no more available today than it ever has been, even though there are thousands of counselors, psychologists, social workers, and even a few psychiatrists who have been very successful counseling people like Joan and her son. The reduced availability of insurance money for counseling is the result of a massive public relations program along with an increasing amount of television, radio, and print advertising that treats brain drugs as panaceas.

An uninformed public that loves the idea of a quick fix easily accepts the idea of mental illness and asks for drugs. The HMOs, like the one Joan uses, view counseling as too slow and too costly. For Joan and her son, it is almost certain that her HMO physicians will provide nothing more than medication. A lot of it will be renewed, changed, or even increased on the phone without a second visit. For people like Joan and her son, brain drugs are the rule, not the exception.

In actual practice, the drug-prescribing psychiatric establishment not only has little interest in mental health, it is moving in the opposite direction: finding more mental illnesses. The *DSM* has moved from I to IV in less than fifty years. The more "mental illnesses" are found, the more drugs are needed to treat them. People who understand this, and who act on the questions I have raised in this book, may be labeled uncooperative and offered on no alternate treatment.

Joan's son is a good example of how pediatricians are being called in to diagnose schoolchildren who don't cooperate in school because they don't like it as having attention deficit disorder (ADD) or attention deficit hyperactive disorder (ADHD). Treating them with narcotic drugs such as Ritalin is only confirming what many psychiatrists and pediatricians already believe: that it's better to use drugs than to try to apply their prestige and clout in the community to the real problem: improving our schools so that students find them enjoyable enough to pay attention and learn in an environment where drugs are not needed.[8] This misguided psychiatric effort has created an epidemic of drug treated "mental illness" in the schools.

It is a paradox that as much as the media is concerned about "crack babies" who are born addicted to cocaine, the same media strongly backs the use of synthetic cocaine, Ritalin, for young children who are unhappy in school. A woman on local public television described being addicted to Ritalin. She ground it up and sniffed it. When she couldn't find a doctor to give her more Ritalin she switched to cocaine and said she didn't have a moment of discomfort. The experience was exactly the same. Is this the drug you want to give to your child?

Once you are started on any brain drug, you may have to stay on that drug for years to keep your brain chemicals "in balance." You will also be warned that the drugs may have to be increased or changed for stronger drugs if their effect is insufficient or seems to wear off. The drugs act on your brain in many ways that are not yet known and may never be known. But what is known is that they often do a lot of harm.

The prescribed drugs may make you feel better, temporarily. In doing so, they are no different from any legal drug you

---

[8]Read my 2001 book, *Every Student Can Succeed,* to see how Glasser Quality Schools eliminate this problem. Further information in this book is in Chapter 14.

may take on your own such as alcohol, nicotine, or caffeine. Or any one of the many illegal drugs you may also use such as marijuana, heroin, cocaine, or methamphetamine, also called speed. The class of drugs called SSRUIs,[9] as exemplified by Prozac, Zoloft, Luvox, and Paxil, have an action, if indeed they do have one, that is very similar to the amphetamines'. I say *if* they have an action, because recent research indicates that they may have no more activity than a placebo. Or that all the hype that surrounds them has brought out a placebo effect that is not actually due to the drugs. See all the research by Irving Kirsch and others cited in footnote 2 in Chapter 1.

They initially give you a lift. That lift may remain for months. But unless you solve the personal problem bothering you, such as an unhappy marriage or a boring, unrewarding job, the initial lift will wear off and you may need a higher dose. But unfortunately, the drug, especially in high doses, may affect your ability to think. Or may affect you physically with bizarre uncontrolled facial and body movements that are symptoms of tardive dyskinesia, which I have already mentioned. There is also the possibility you will become addicted to the drugs, which is hardly a desirable outcome.

In what seems to me to be inconceivable naiveté, the establishment psychiatrists believe that these powerful drugs cannot damage your brain even though there is overwhelming evidence to the contrary. If you don't feel good on the drugs, they will urge you, force you, if they have legal control over you, to keep taking them because, regardless of how bad you may feel on the drugs, they believe you are now on the path to "true" mental health: a well-balanced brain chemistry. With this newfound "chemical

[9]Selective serotonin reuptake inhibitors that purport to lift you psychologically by increasing the level of the brain chemical serotonin in your brain.

mental health," you will stop being depressed, schizophrenic, bipolar, or anxiety ridden, even though the problems in your life that cause your unhappiness have not been solved.

But if they are wrong, and there are many highly respected psychiatrists and psychologists who believe they are completely wrong,[10] it should reassure you to know there is a small group of psychiatrists like myself who don't believe in mental illness. Our minority voice continues to struggle to protect you and is being heard in many other books besides this one (see the appendix). Part of what the ICSPP is doing now to protect you is to launch a campaign to stop the advertising of strong brain drugs directly to you on television and in the newspapers.

In the Treatment of Depression, the Placebo Effect of Sugar Pills Is Strong Evidence That Unhappiness, Not Mental Illness, Is the Cause of the Symptoms

If depression were caused by a chemical imbalance in the brain chemistry, then it should not be relieved by a sugar pill. Yet evidence is mounting that given with care, conviction, and in a context including time spent with the doctor who gives it, the sugar pill works better than antidepressant drugs such as Prozac, Paxil, and Zoloft,[11] Interesting also, is that when brain activity scans are used to confirm the "positive effect" of these

---

[10]The major organization that is fighting organic psychiatry and its beliefs is the International Center for the Study of Psychiatry and Psychology, or ICSPP, founded by Peter Breggin, M.D. See the Appendix of this book for more information.

[11] Shankar Vedantam, "Sugar Pills Fight Depression," *Washington Post*, May 7, 2002, which cites extensive research to support this conclusion.

drugs on the brain, the brain scans of people on sugar pills show the same or even greater activity than the real drugs. Placebos also may be more effective because they do not have the adverse side effects of real medication.

For example, suppose you are depressed because you are unhappy with a relationship in your life as Joan described herself in this chapter. You go to your doctor; he listens to your story, prescribes a medication, and then tells you with some conviction (because he believes he is telling you the truth), "This is the latest medication for depression. It has helped many of my patients. I think it has a good chance to help you. It takes a while to work but please call me in a week and tell me how you are doing." With this much attention, patients, most of whom want to please an attentive doctor, will report improvement. What is interesting is that, after the study was finished, some of the patients who reported strong positive placebo effects were told that they had received the sugar pill. They relapsed immediately. The placebo effect had been shattered by this revelation.

Studies like this on the effectiveness of placebos have been conducted for centuries and always turn out the same. What is vastly different about studies involving brain drugs is the huge media support of reports of their effectiveness. Billions of dollars are being made with these drugs; there are no corporate profits to be made from counseling or programs to improve mental health.

Like a psychiatric cancer, the false belief that only a drug can cure mental illness, has invaded our whole society, reducing the practice of counseling and, by convincing you that you are mentally ill, standing directly between you and mental health. What is so ironic is that the HMOs, which mostly control your access to psychiatric care, have climbed on the mental illness, brain-drug bandwagon because they see psychotherapy as more expensive than drugs. But drugs are not cheap.

To put a patient on Paxil for a year, 365 pills at three dollars each, costs over a thousand dollars and brings little certainty that his "disease" will be improved to the point where he has no need for further drugs. To the contrary, once the drugs are started, their use tends to escalate, especially since many of them are addictive. The Choice Theory Focus Groups, which I will describe later in this book that teach you how to live your life more effectively, will cost you nothing. Since you can keep using what you have learned, the result will be much longer lasting. This will be a win-win-win opportunity for the patients, the psychiatrist, and the HMO.

Suppose an HMO were to get interested in a mental health program. They could easily find one or two psychiatrists who, besides prescribing drugs, would also devote a few hours a week to setting up and supporting the program I will describe later in this book. Considering how many HMOs there are, I don't believe there would be much trouble finding enough psychiatrists to tackle this interesting challenge.

A further benefit to the HMO of a mental health program is that it would provide their medical doctors with another option for the many patients they see every day who complain of pain or discomfort for which no physical cause can be found. Even after the expensive MRIs and CAT scans are done, plus a lot of doctor time invested, patients suffering pain keep returning to their doctors.[12] I recognize that these patients will be highly skeptical of any mental health program, protesting that what they need is more physical care and better medication and that their mental health is fine.

Yet for the nine years that I was the psychiatrist for the Los Angeles Orthopedic Hospital, during which I worked with over fifty chronic pain patients, no cause for their pain was

[12]See my book *Fibromyalgia* referenced in Chapter 14.

ever determined, and I found them remarkably open to the idea that the pain might not be medically treatable. I gave them time and created a relationship with them and the results in many cases were very good.

I did not have access to choice theory at the time. In those years I wasn't thinking in these terms. But I can see now how effective a program that gave groups usable mental health information plus time and attention could have been. Such a program would be active, with a lot of give-and-take, not just lecturing and listening. It would also encourage attendance by using a friendly, we'd-like-to-see-you-again approach. There would be none of the long, lonely wait and then the hurry-up encounters they run into now when they finally see a doctor. As much as Choice Theory Focus Groups would benefit the patients, it would also benefit their doctors who don't know what to do with their patients now.

Get the Mental Health Associations Involved

In this chapter I have explained the need for a better mental health program and have begun to suggest what I think it might look like. But what I would like to stress now is that the natural place for this program is in the many Mental Health Associations that now span our country. Using this model, they could begin doing what their name implies: improve the mental health of their community at no cost to the people who enroll.

But you have to realize that this is my vision for the future. I see you using this book to improve your own mental health right now or to help friends or family members to improve theirs. Eventually, getting together as a group to do this could be very effective. If you've read this far, you've already begun the journey.

three

Unhappiness Is the Cause of Your Symptoms

Unhappiness is best described as a time and place in our lives when our life is not the way we'd like it to be. We can be in this unhappy place for a moment or many years, but as soon as we realize we are in it, we will do whatever we can to get out of it. If our unhappiness continues for months, symptoms such as depression, anxiety, mania, panic, headaches, even symptoms associated with what is called schizophrenia will appear. Because we are creative, there is always the possibility that a new symptom will make its appearance. That's one of the reasons the *DSM* has grown so quickly to the large volume it now is. Why and how this happens will be explained in a later chapter.

Assuming we are physically healthy and have sufficient food and shelter, we experience more unhappiness in our marriages than anywhere else. The initial symptoms will be anger and depression. The depression will last longer and longer if we don't do anything to improve our mental health. If it continues, additional symptoms such as fatigue, headache, listlessness, difficulty sleeping, and loss or gain of appetite may add to or replace depression. It is virtually impossible to be unhappy for longer than five or six weeks and remain symptom free.

It is safe to say that no one can avoid unhappiness. All we can do is try to understand what's wrong and, from this understanding, try to figure out what to do to become happy again. If we can regain our mental health, our symptoms will disappear. The question I will answer in this book is: How does unhappiness lead to symptoms and what can we do to become happy again and symptom free?

To begin, I'd like to return to Joan, the unhappily married woman I introduced briefly in Chapter 1 and described in more detail in Chapter 2. Through a mutual acquaintance, she'd heard about me and my interest in helping people without drugs, and called for an appointment. Hers was an all too familiar story. She explained that the psychiatrist she had seen was more interested in her symptoms, her diagnosis, and in pushing medication than he was in the problems she was having in her marriage.

Joan then said, "I have a very unhappy marriage and I'm depressed. I understand you don't use drugs but now you tell me you no longer counsel."

I said, "Joan, I've done a lot of marriage counseling and you certainly fit the profile of most of the women I've counseled. But for the time being, I've stopped counseling. I am working on a different and longer-lasting approach: teaching people like you to improve their mental health. If you can learn how to do this, you may never be depressed again for more than a short time in your marriage or any other part of your life.

"You mean as miserable and depressed as I am, I'm not suffering from a mental illness? The doctor who diagnosed it told me that I was clinically depressed and suffering from a chemical imbalance in my brain. He read me a paragraph on clinical depression from a big red book of mental illnesses that described me to a tee."

"I believe the doctor's wrong and the book's wrong. The description of your suffering that he read you is correct but there is nothing wrong with your brain. A normal brain is perfectly capable of doing to you what you are experiencing and worse. Joan, no matter what that doctor told you, you are not mentally ill."

"But if that doctor is wrong, then what am I suffering from? I'm depressed all the time. Believe me, I'm not imagining what I'm feeling."

"I believe you completely. I'm sure you're depressed but what you're actually suffering from is the most common of all human complaints: you're unhappy. When you're unhappy, you're not as mentally healthy as you'd like to be, but you are not mentally ill. The way I see it, happiness is mental health and from what you've told me on the phone, I'm sure both your husband and your son are unhappy, too. There are thousands unhappy people like you for each person who suffers from a mental illness. You know, someone who has something physically wrong with his brain like Parkinson's or epilepsy."

When I explained that, Joan was quiet for a while. Then she said, "Well, it's good to know that nothing's wrong with my brain. And you're right about me being unhappy. I've been unhappy for a long time. But I still don't see how something as mundane as unhappiness can cause how I feel. Half the world's unhappy."

"I'm not sure it's that many but there's no shortage of unhappy people. What you feel, your depression, is the most common way that unhappy people deal with their unhappiness. As soon as you improve your mental health, you'll stop being depressed."

"Sounds good. How do I do it?"

"I'll teach you. It's kind of like improving your physical health. There are millions of overweight couch potatoes who aren't sick but they know they aren't healthy. Most of them know how to get back to physical health, they have to exercise and eat less. The difference is, if you're not mentally healthy, you don't know what to do. When you learn what I teach, you'll know."

"Are you saying you're not going to counsel me, you're going to teach me."

"That's exactly what I'm saying. I've already started. I've begun to teach you the first important lesson."

"What's that?"

"That you're not mentally ill. There's nothing wrong with your brain. If I can get you to consider that, we're on our way."

"Well, if all you're asking is I consider that, no problem. But say you'll teach me how to be mentally healthier than I am. How will I know if I've learned something?"

"Like I said, you'll begin to feel happy again. You know what happiness is. You haven't been unhappy all your life."

"Okay, I'm here, let's get started. What do you want me to do?"

"For now all I want you to do is talk to me for a while longer and then go home and read some material on mental health that I've already written. If the material makes sense to you, we'll go from there. What I'm doing is starting to write a book on mental health. If it's okay with you I'll record the rest of this session. If it comes out the way we both want it to, it'll be in the book. I've already written a chapter. In it I cover what we've already started to talk about and a lot more."[1]

"Is this some kind of self-help book?

---

[1] That chapter was Chapter 2, which you have just read. At this point, the Foreword, the Preface, and Chapter 1 had not yet been written.

"It is. But it's more than that."

"What do you mean?"

"Eventually, I plan for it to be a help-each-other book. If you can get together with a small group who've read the book, you can help and support each other. You can still use it to help yourself or a member of your family. But my idea is that instead of each trying to help yourselves, you'll all meet together in what I call a Choice Theory Focus Group. You'll discuss what you've read and how you can help each other, as well as help yourself, to become mentally healthier. In your case, because you're the first one I've shared this material with, I'll help you to get started on your own and see how far we can go. But if we can get a group started, you won't need me anymore. There's nothing in this book that anyone who reads books or newspapers can't understand. Supporting each other as you put the material in the book to work in your lives is the goal of the focus group. Can you picture what I mean?"

"Do you think you could get such a group started?"

"People read my books and write me that they've gotten help from them. I figure they may be interested in a book like this. If you read it and get interested, that'll be a good sign."

"How many people would be in the group?"

"That would be up to the group members. Just not so big that you can't get to know each other and have a chance to express yourselves. I can see groups of about five to fifteen people. They don't even have to know each other. My Institute has an online chat room. We can tell them to read the book, then click on to the Choice Theory Focus Group and talk to each other. Sometimes I could be on or my wife. It can all be posted on our Web site. If it's a good idea, it'll get going. But I'm hoping you'll help me get a group started by reading the book."

Joan laughed and said, "You see us as a cage full of guinea pigs searching for happiness?"

"I guess you could say that, sure. When they use guinea pigs for research, a lot of them get help. For us, the rest of the world can be a control group. Guinea pigs, that's good. I'm going to enjoy working with you."

"Is there any risk for me in what you're asking me to do?"

"None that I know of. All you'll do is talk with me to get things started. Then, if you get some benefit from what you begin to do, we'll ask others to read what I've written and enlarge the group. No one will be obligated to do anything. Drop out any time. And because the group members quickly take over and start doing it themselves, it won't cost anything. As we talk I record what we're saying. In a short time we'll have most of the manuscript."

"How often do you and I have to meet?"

"I can see you and I getting together frequently in the next few weeks. If we get a focus group going, I can see them getting together once or twice a month. My wife and I started a group like this for six young adults who were suffering from rheumatoid arthritis. My theory was that improving their mental health would improve their physical symptoms, and they reported a lot of improvement. We met once a month for sixteen months and it seemed to work. This was one of my many experiences with teaching mental health that led me to start working on this book."

"But what if it doesn't work?"

"Well, there's no guarantee. But wait, there is a guarantee. Reading the material and seeing me for a while can't harm you or anyone else. It worked before with the young people suffering from arthritis without a book, as we focused on mental health. Besides, I wouldn't go to all this effort if I didn't think it would work."

"One more question. Suppose I like the book and the focus sessions but I still want to get counseled, can I do both?"

"Of course you can. You can even take the Prozac the psychiatrist you saw prescribed. What you do outside the sessions is up to you. But to avoid getting confused, I'd recommend you attend a few sessions before you get involved in anything else. I'm ready to start right away. We can start tomorrow if you read what I've already written, tonight."

The Chinese say a long journey starts with a single step. Joan and I had just taken ours. What Joan and I and the others talked about is simulated. But what goes on in these conversations is based on my many years of experience and working recently with the arthritis sufferers. The sessions, even this one, do not need to be reported verbatim. I have removed most of the small talk and repetition that's standard to a friendly, supportive group interacting about a subject they're all interested in. When the groups got bigger, the sessions became longer and more detailed.

four

The First Choice Theory Focus Group Session: Choosing Your Symptoms

Joan, I'm glad you're here. I was so happy when you called to tell me you liked the two chapters I gave you. Before we start, do you have any questions?"

"No, go ahead. Believe me, I agree with what you wrote. I'm plenty unhappy but I don't see myself as mentally ill. I don't see my son as mentally ill and needing a drug, either. If he was happy in that class, I'd really worry about his brain."

"Okay, then let's talk about happiness. Could you tell me when you were last happy? I don't mean for a month or two. I mean for a period of time, a couple of years, maybe longer."

"That's easy; it was for six years, from the time I met Barry until the baby came. I'd never been so happy as I was during those years."

"Can you remember how you treated Barry when you were happy?"

"I don't know, I really didn't pay any attention. We never argued, we never fought; I got a lot of love and I gave him a lot of love. That's what's wrong, now. I don't get much love and I don't feel like giving any to him."

"If it's all right with you, I'm going to concentrate on what you did or what you're doing. I'm not going to ask much about what anyone else does, like Barry or your son. Do you find much fault with Barry?"

"Why aren't you going to talk about him? We're married. Isn't what he does as important as what I do? I hate what he's doing to me."

"That's a good question, I guess I should have answered it before. The reason I'm going to focus on what you do is because what you do is all you can control. There's no sense talking about what anyone else does, because you can't control what they do. No matter how hard you try, the only person's behavior you can control is your own. Can he control you? Can he make you do anything you really don't want to do?"

"He sure tries."

"I'm sure both of you try very hard to control the other. But, please, let's get back to my question. It'll help you to understand why I only want to talk about you. Did you find fault with Barry when you were happy?"

"Of course not. We were happy. We had a few arguments but one of us always gave in and it was over."

"How about now? Do you find fault with him now?"

"That's all I do is find fault with him. But believe me, he finds fault with me, too."

"You don't give in like you used to?"

"Why should I? He doesn't give in to me."

"Joan, if you really want me to answer that question, I will."

"What question? What are you talking about?"

"You asked me, why should I give in? Do you really want me to answer that question?"

"I did say that, didn't I? But I didn't expect you to answer it. You have an answer for that question?"

"I do, but only if you still love him. I'm guessing you do, right?"

"Of course I still love him, If I didn't, I wouldn't be so miserable. I'd just like him to be the way he used to be."

"I thought so. Good. A lot of unhappy wives still love their husbands. But if you do, shouldn't you consider giving in once in a while like you used to, even if he doesn't. It's not that if you give in, you've surrendered for good. You haven't lost control over your own behavior. You'll always have control over what you do."

"But I don't want to give in. I can't. It's not fair."

"You gave in when you were happy; was it fair then?"

"You're getting me all confused. I don't want to give in. If I do, he'll win and I'll lose. He ought to give in."

"Do you think he's going to give in?"

"I doubt it. He's stubborn. But he should."

"Can you make him give in?"

"I've been trying. He hasn't so far."

"But you keep trying. When was the last time you tried? This morning?"

"It wasn't this morning; this morning was okay. It was last night. He wanted to make love and I didn't want to."

"You didn't want to make love or you didn't want to give in? If you didn't want to, just say so. You don't have to tell me why."

"But I did want to. I wanted to a lot. But if I did, I'd be giving in."

"Look, please, I'm not criticizing you, Joan. Whether you do something or don't do something is not my concern. What's important is what I'm trying to teach you. Last night you had a choice about doing something you wanted to do and you chose not to. It's your life, choose what you want, but I want you to realize everything you do is your choice."

"Okay, I chose to turn him down. Right now it seems stupid. At the time, it seemed like the thing to do."

"I can't teach you about stupid or smart, right or wrong, good or bad. That's all up to you. But right now, while you're unhappy, you've been choosing to do a lot of things and I don't think you realize they're all choices. You could choose to do something else that's even worse than what you did."

"I turned him down when he was trying to be loving. What could a wife do to a husband that'd be worse than that?"

"You could've blamed him, criticized him, complained about him, started to nag him about something else. You could also have gotten more depressed and been unable to sleep. I haven't explained it yet, but I believe most of your symptoms are choices, too."

"I kind of gathered that from what I read in those chapters. But believe me, they don't feel like choices, they feel like they're happening to me."

"I know that's how they feel and don't worry about it. I cover it in later chapters. But how you deal with Barry when you're unhappy, that's what we have to talk about now. When he asked you to make love, you were angry, weren't you? You kind of stopped being depressed and got angry."

"You're right, I did. Actually, I'm angry a lot but I keep it in because he gets angrier and it leads to a fight. When I get depressed he doesn't get angry and it's better."

"Even though you kept it in, who do you think was responsible for your anger?"

"That's easy. He was. He was making me angry. If he wants to make love, he should treat me better. And not just because he wants sex. I want good treatment all the time."

"I agree. You'd both be better off if you treated each other better. But what I want to teach you now is very hard to learn. Joan, Barry wasn't making you angry. He can't make you feel anything or do anything. No one can. I sense you're getting a little angry now at what I'm saying."

"I am. What you're saying doesn't make sense."

"What I'm trying to teach you is a new kind of sense. Relax and give me a chance. If you want to be happier than you are now, you have to accept that, when you turned him down last night, every word that came out of your mouth was chosen. And along with those words comes a tone of voice, an expression on your face, every gesture with your hands and all your body language. It's all chosen. You could have said, 'Okay, Barry, that's the best suggestion you've made in weeks' instead of turning him down. If you'd said that, all the other stuff, your tone of voice, the expression on your face, it would have all been different. Very different if you'd enjoyed yourself."

"I felt like saying that. I told you that. Those good times we had when we were first married weren't perfect but that's what we did. We sure don't do it anymore."

"It's what you chose to do then and what you're not choosing to do now. I know it seems like I'm harping on the words choice and choosing, but those words are the heart and soul of the book I'm writing."

"I see what you're driving at. This is hard. If this is what I have to learn to become mentally healthy it's going to take some time."

"There's no hurry. Take all the time you want. But keep in mind it's painful time and you have to suffer the pain that goes with it. When you chose to turn him down you felt like it was the right thing to do, didn't you."

"What do you mean, I chose to turn him down?" She paused for a moment and then said, "I did, didn't I?"

"Joan, I know this is hard, it's going to take time and I'm going to start using the words choosing and choice a lot to kind of remind you. But like I said, when you think you're right, it's especially hard and you tend to get angry. That you're making a choice doesn't enter your mind. We pay a big price for the feeling that we're right. You've been paying that price for a long time."

"I've got something here to learn, don't I? [I nodded.] I guess I have been choosing to turn my back on him. As soon as I do, I feel depressed."

"Covering up your anger with depression is no picnic. I'd like to teach you a little trick to help you learn that you're choosing everything you do."

"I could use a trick or two. Believe me, every bone in my body tells me Barry's causing my pain."

"Okay, here's the trick. No matter how you feel, as much as you can, before you do or say anything, say to yourself, 'I'm choosing to say this or do this.' The next time you decide not to make love, say to yourself, 'I'm choosing not to.' When you say that, it's harder to blame him. I use this little trick myself and I've taught it to other people. Believe me, it really works."

"You're trying to get me to stop blaming him."

"I'm not trying to get you to do anything. I'm trying to teach you that whatever you do is a choice. Being here with me is a choice. You could choose to leave and never come back. Accepting that everything you do is a choice is the cornerstone of mental health. This is teaching, it isn't therapy. I'm not trying to get you to change your life. I'm trying to teach you something that, if your life needs changing, will help you to change it. Remember what I said in the chapter you read about mentally healthy people and happiness."

"I remember. You said that they aren't happy all the time. Believe me, I fit that description."

"But I also said that when mentally healthy people aren't happy, they can usually figure out what to do about it."

"I wondered about that. I've been unhappy for years and I never really seemed to know what to do about it. I guess I'm not all that mentally healthy, am I?"

"You'll be a lot more mentally healthy if you can learn that all you do or say is chosen and that you almost always have other choices, often better ones."

Joan stopped talking for a while and just stared out the window. Then she said, "Is that all for today? This is plenty for me to think about."

"If you've got the time, I need another few minutes. There's one more thing I'd like to teach you. It'll help you to recognize you're making a choice if you forget to say to yourself, 'I'm making a choice.' You've been unhappy for a long time and you told me you're depressed a lot. Now I'd like to teach you how to recognize that when you're depressed, you're choosing that, too."

"Keep trying, but like I said, when I'm depressed, that's how I feel. I'm not choosing it. No one would choose to feel that way."

"Okay, I won't argue. But tell me, when you say, 'I'm depressed,' what do you mean?"

"I mean I'm upset. I feel rotten, hopeless, helpless, discouraged. . . . That's it, I feel so helpless. That's a good way to put it. I hate that helpless feeling."

"Helpless, that is a good way to put it. Would you like to feel less helpless when you think about your marriage and how your son's acting in school?"

"Sure, get me a new husband and a new kid. Okay, I'm sorry."

"No problem; if you weren't feeling a little better you wouldn't joke about it. You're not depressing as much as you were when you came in."

"Depressing? What are you talking about? I was depressed when I came in. A psychiatrist before I saw you told me I was suffering from clinical depression."

"Would you consider not using those words?"

"What words?"

"*Depressed* and *depression*."

"Why?"

"Because they're not accurate. Mentally healthy people are accurate when they talk about their feelings. There's nothing in those words, *depressed* and *depression*, that convey the message that you're choosing how you feel. But if you start saying, 'I'm depressing' or 'I'm choosing to depress,' as soon as those words come out of your mouth, you introduce the idea of choice and choosing. If you're moving along down the sidewalk on your feet, you say, 'I'm walking.' You don't say, 'I'm walked.' We're talking now, not, we are talked. Do you see the difference? Depressed is like the way they describe the economy. What can you do about the economy? Depression is a weather pattern they talk about on the Weather Channel. What can you do about the weather?"

"How about if I'm angry?"

"Then you're angering or choosing to anger. You don't have to say them out loud, but the more you do, the better it is. If people pick up on it and ask why, explain it to them. It's good practice. As soon as I say, depressing, or, I'm choosing to depress, I start thinking there must be a better choice. I turn my attention to what I can do about it and pretty soon I stop depressing."

"I keep telling Barry I get depressed whenever we talk about Josh, our son."

"But then you're intimating that it's Barry's fault and Barry needs to do something about it."

"So you're asking me to say 'I'm depressing' about Josh or 'I'm choosing to depress' about him."

"Why not? It's the truth. Blaming Barry doesn't help Josh. It certainly doesn't help your marriage, does it? Explain it to Barry. It's not just you. He's been blaming you just as much as you've been blaming him. Blaming doesn't solve anything. It makes things worse. Choosing leads to solving problems. One of these days Barry may want to come in here with you. I can teach the two of you just as easily as one. I'll get this chapter, what we're taping right now, to you in a day or two. Let him read it and all the other material I gave you and see if they make any sense to him."

"That'll be the day."

"It will, won't it."

"You really believe this stuff can be learned and Barry can learn it along with me, don't you?"

"Of course I do. I wouldn't teach it if I didn't believe you could both learn it. You have nothing to lose by letting Barry in on what you're learning. It's a lot better than fighting with him."

"Just because I'm choosing to do something doesn't mean I'm going to make a better choice."

"When people realize they're making unhappy choices, they usually want to make better ones. As long as you accept that you're choosing to depress, it's hard not to try to make a better choice. Being *depressed* is self-defeating. Suffering from clinical depression is even worse. It increases the helplessness you feel. The last thing that you believe if you think you're mentally ill is *I can do something about it.*"

"What do you mean?"

"I mean that as long as you continue to depress and wait for Barry to change, you give him control over your happiness. You can't feel better until he changes. Since you can't control his behavior and he can't control yours, you're both stuck in your misery until at least one of you starts to make some better choices. I'm hoping it'll be both of you. Ready to quit for the day?"

"When can we get together again?"

"Some time about two weeks from now. Feel free to bring Barry along if he wants to come. I'll be adding some material to this chapter that I think you'll find interesting. Bring them all along when you come back. We can refer to them; it could help. I don't know how long this will take, but when we're finished you'll have the whole book. See you in about two weeks. I'll give you a call when I'm ready. Remember, good choices are those that bring us closer to all the people we want to be close to. Bad choices tend to separate us from those people."

Mentally healthy people don't accept that suffering is an unavoidable part of life. They see their suffering as a warning that right now *I'm very unsatisfied* with one or more of the important relationships in my life. This could be a mate, child or children, parent, friends, students (if you teach), coworkers, or boss. There may be others but, if you are on good terms with all of these, you're about as mentally healthy as it's possible to be.

Mentally healthy people pay attention to that warning. They know that if they are to get back to being happy, they must take a good look at what they, not the others, are choosing to do in all their close relationships. That is because they're aware that they can't change what anyone else does; they can only change what they're doing. Most people deal with unhappiness the way Joan did: making painful, symptomatic choices and blaming how she felt on the other person in the relationship. But there will be times when Joan might not be sure. She might ask herself, "Is there a way to tell if I'm making a bad choice before I go ahead and make it?"

The answer to that question will be covered in some of the remaining chapters but it can be accurately summarized here: Good choices are those that bring us closer to the people we want to be close to. Bad choices tend to separate us from those people. That may seem to you to be an inadequate answer because you may ask, how will I know if it's good or bad before I actually make the choice? A few chapters from now, I'll be able to answer that question in detail.

But for now, *I can only tell you, you'll know.* You'll feel a sudden discomfort, the same discomfort you feel when you say to someone you want to get along well with, "I know I shouldn't tell you this, but . . ." That discomfort is a warning that, if you say what's on your mind, you're risking your relationship with that person. The smart thing to do is pay attention to that warning and keep your mouth shut. Good relationships are the core of mental health and happiness. Whatever little thrill you'd get from conveying that bit of news isn't worth jeopardizing your relationship. If, however, you're expressing that tidbit with the intent to hurt, you really need to pay attention to this book.

These ideas are new and hard to accept. Joan may read this and still say to herself, "What it gets down to is he's saying I need to change, not Barry. But why me? It's not my fault, it's his. I don't want to. It's not fair."

I answer that often asked fairness question in mental health terms by saying, "Joan, it doesn't make any difference whose fault it is or whether it's fair or unfair. There is no happiness in an eye for an eye or a tooth for a tooth."

Mentally healthy people avoid being self-righteous. They look for a better choice. A choice that isn't better for one over the other but a choice that's good for both. The best choice is the one that helps their relationship, not the one that helps one or the other individually. If changing what you're doing will help the relationship, go with that, no matter whose fault it is or how unfair it is. Keep your eyes on the doughnut; there's no nourishment in the hole.

For the rest of this book I will explain things just as I've been explaining them so far: some by direct conversational teaching, the rest by providing of information as I'm doing now. If these ideas make sense to you, as I'm hoping they will, you'll be on the way to better mental health. If you're not willing to focus on changing your life, and to stop trying to change other people's lives, you'll continue to be unhappy.

Now that I've explained choice and change, I'd like to explain a major difference between being less than mentally healthy and being mentally ill, for example, the difference between depressing and suffering from Alzheimer's disease. When your brain is damaged or diseased, it loses its flexibility; your choices become limited or cease to exist. There is nothing wrong with Joan's brain. If she can't accept that she has choices, it's almost as if she's choosing to put her brain in a straitjacket. She may be stubborn or foolish but she's not mentally ill.

As much as we have choices, there are some unhappy situations we can do little about. For example, if we exist in hopeless poverty, suffer from an incurable disease, or are forced to endure life under a tyrant who robs us of our freedom, it is hard to be happy. But for the readers of this book, the vast majority of your unhappiness will have nothing to do with those extreme situations. If you're unhappy, it's because you're not getting along the way you want to with the important people in your life.

When a mentally healthy person tells you he's unhappy, he follows it up by saying, "Here's what I plan to do about how I'm dealing with this unhappy relationship." Or asks, "Have you any suggestions that might help?" Because he approaches his unhappiness without blaming anyone else, people want to make suggestions and even give him some help.

If there is mutual love and respect in your relationships, you can endure misfortune, even calamity, and still be mentally healthy. Morrie, the dying professor, chronicled in the best-selling book, *Tuesdays with Morrie*,[1] was in such a situation and remained mentally healthy. It was his optimism, cheerfulness, and caring, as much as his wisdom, that attracted readers to that book.

Muhammad Ali suffered severely because he would not surrender his belief that killing people who had no hostility toward him was wrong. Because of his conviction, he refused to serve in Vietnam, but in the end he gained respect.

These are examples of the ways mentally healthy people deal

[1]Mitch Albom, *Tuesdays with Morrie* (New York: Doubleday, 1997).

with adversity. They refuse to submit to unhappiness. They struggle against it and do what they can for themselves and don't depend on what others can do for them. Even in the face of criticism and rejection, they continue to love and respect themselves. You may be thinking this is too simplistic. How can such a common, easily understood word such as unhappiness possibly describe all the misery and mayhem that exists all around us? My answer is: Unhappiness isn't simple. As I will continue to explain, it can appear in our lives in many different forms and lead to a myriad of feelings, thoughts, and behaviors that puzzle, frighten, and disturb us. Severe unhappiness can lead to bipolar disorder, schizophrenia, and chronic, excruciating pain with no pathology to explain it, as in the condition called fibromyalgia.

A more accurate title for the *DSM-IV* would be *The Big Red Book of Unhappiness*. No matter how we experience it, almost every symptom can be traced back to its origin: relationships that lack love, respect, or both. By making choices that help us stay connected with each other, the unhappiness can be overcome. Caring and respecting, never controlling, are the cornerstones of mental health.

The Difference between Happiness and Pleasure

A key component of mental health is knowing the difference between happiness and pleasure. As much as these two experiences may seem to be the same—both feel very good—they are very different. Happiness is feeling good because you are choosing to behave in ways that keep you close, or get you closer, to the important people in your life. Pleasure, often associated with addicting drugs, gambling, or casual sex, may, for short periods, feel better than happiness. But we should not be fooled; pleasure is not happiness. It is another experience altogether.

For example, many bars have a "happy" hour starting late

Friday afternoon when the work week is over. If you are in the bar, you'll see a lot of men and woman talking to each other and laughing. Sexuality is in the air. They tend to listen to each other and show interest in each other, more than many do at home and at work during the rest of the week. All this is augmented by the alcohol, the most common pleasure drug in the world. Take away the alcohol and there'd be a lot less camaraderie.

But take away the camaraderie and there'd be very little laughter. People drinking by themselves don't laugh. What is going on in the happy hour is social drinking, the pleasure of relaxing, showing off, flirting, and watching others do the same with friends or friendly strangers. Alcohol, often labeled a depressant, does not depress these people at all. It helps them feel at ease, loosens their tongues, and releases their inhibitions.

But your ability to enjoy these happy hours depends on accepting the unwritten rule most participants follow: *What I do or say starts and ends here in the bar.* Certainly, affairs may start and marriages may end, but if this happened too often the safety of the happy hour would be threatened. As long as everyone plays by the rules, it's fun to show off and flirt.

But for some people at the bar on those afternoons, it is the alcohol, not the company, that makes the drinking so pleasurable. If these people start to flirt or argue, they may not play by the rules, but may go too far and suffer rejection. The others in the room soon recognize that their behavior is being motivated by the alcohol and are cool to their efforts to socialize. Sensing this rejection, they get hostile and turn more to the sure pleasure of alcohol and drink heavily, much more than the social drinkers, and stay later, often until after most of the social drinkers have left. Many get into their cars too drunk to drive.

Heavy drinkers, habitual gamblers, and sexual addicts who depend on getting pleasure out of their addictions, not from their relationships, are not happy; *they do not have, or, if they are addicted, seem not to want the satisfying relationships most of us need to be happy*. But when you are unhappy and go to a psychiatrist who prescribes brain drugs, it is very easy to mistake the pleasure you may be getting from the drugs for happiness.

It is especially easy to make this mistake if you are prescribed a brain drug and told not to worry about addiction because this drug will cure your mental illness. If you take such a drug and are counseled to use your newfound energy and better feelings to solve your relationship problems and find happiness, the drug will have served its purpose.

But that rarely happens. The drugs are prescribed with the idea that feeling better is all you need. But it isn't all you need. It's even worse than all you need, because you mistake the pleasure of the drug for happiness. You may feel better but you are just as lonely as you would be if you were in a bar drinking alone. That loneliness will soon overcome the pleasure the drug provides. There is no happiness drug, legal or illegal, that brings people closer together.

If you are given the brain drug and not taught that what you are feeling is not happiness or mental health, very often you are worse off than before you took it. Brain drugs such as Prozac, Zoloft, Luvox, and Paxil are so popular because this particular class of drugs has (or may feel as if it has) an amphetamine, or "speed," effect on the brain. But your brain under the influence of the drug is no longer normal and less able to be flexible enough for you to make the choices you need to make to be happy.

Your brain, programmed by millions of years of evolution without drugs, deals with drugs as unwanted invaders and fights to rid itself of their effect so it can return to normal functioning. As it does this, you actually experience a chemical imbalance. But as many of you know, even under the influence of alcohol you still have a good idea of what you're doing. In this abnormal situation neither the drugged part of the brain nor the normal part is dominant. The pleasure of taking the drug may disappear as your brain fights against the drug and, to regain the pleasure, you may require more of the drug. When that happens you are addicted and your brain is no longer able to function normally.

While this is happening you can't feel good without more of the drug and you are unable to help yourself. When the drug no longer works, instead of weaning you slowly off of it and teaching you what you need to know to improve your mental health, your doctor is very likely to "correct" the artificial imbalance in your brain with more of the same drug or a different one.

If more of the same drug or a new one "rebalances" your brain and gets you back the pleasure you lost, you are now hooked on a legal addicting drug. Many of you reading this book have been on that familiar increase-or-change-the-brain-drug merry-go-round or have seen a loved one on it. It was seeing her brother-in-law on it that prompted the Australian woman to write the letter that began the second chapter of this book.

You can only find happiness in choosing to change the way you relate to the important people in your life; however, once you feel good on a drug, prescribed or self-prescribed, the incentive to do anything other than take the drug and enjoy its pleasure is diminished or removed.

An alcoholic is unhappy when he's not drinking because the pleasure of alcohol, always on his mind, is more important to him than any happiness he believes he can get from the relationships in his life. The more he drinks, the more he destroys what relationships he may have and, in doing so, destroys any chance he has for happiness. The less happiness he has, the more he'll be motivated to drink: the familiar downward spiral of alcoholism.

Most people who work with alcoholics don't believe they can return to social drinking. The immediate pleasure of alcohol is too strongly entrenched in their brains for them to restrain their drinking long enough to build or rebuild relationships. But strong habits, even addictions, can be overcome if the addict can find happiness *without* the drug.

This is a big *if,* but half the millions of alcoholics who regularly attend AA meetings, looking for happiness in the companionship of other AA members, find that happiness and stop drinking. But to keep what they have found, they need to continue to build strong relationships. It's more difficult with legal addicting brain drugs; there is no program comparable to AA's to help them stay off these addicting drugs. It is my hope that the Choice Theory Focus Groups, like the one I have already started to develop with Joan, could become such a program for people addicted to any drug, legal or illegal. I also see no conflict between AA and a focus group. One could support the other.

A friend of mine, an elementary school principal, had to deal with a mother who came to school begging him to let her take her child's Ritalin home, saying her home supply had run out. A little investigation uncovered the fact that she was addicted to her child's medication. Wouldn't it be helpful for the principal who discovered this situation to be able to refer her to a Choice Theory Focus Group? He could lend her this book to read and, if she believed it could help her, urge her to join a focus group.

Don't rush through this book if you are unhappy. Try to find someone who will read it with you or at least talk to you about it. The effort of explaining what you've been learning will help you to get a better understanding of the ideas. Keep in mind that unhappiness, not mental illness, is your problem. You can find happiness if you understand the importance of choice in your life.

five

We Have Learned to Destroy Our Own Happiness

They lived happily ever after" is the storybook ending to books and movies. It implies they'll continue to love each other as much as they did when they were shown sharing that final hug and kiss. But for most of us, happily ever after turns out to be the most elusive assignment we'll ever tackle. After the early passion cools, unhappy choices become easier to make than happy ones.

Without any idea of what they are doing, our parents, grandparents, siblings, teachers, and the people we've worked for have given us lesson after lesson in that art of making unhappy choices. Through observing these people and listening to them, we have learned those lessons well. By the time we are grown, we actually know more about how to make unhappy choices as we deal with the important people in our lives, especially our husbands and wives, than to make happy ones. As we do, we move further and further away from mental health.

These unhappy choices destroy more than half our marriages. Many couples divorce; others hang grimly together, but there's little joy left in those sullen relationships. From these failures, we are overrun with unhappy children. On top of what we fail to do as parents, the schools don't shirk their contribution. Unless students have a special talent like music or athletics, over half the children who attend our schools are unhappy because they don't choose to work hard enough to earn a legitimate A or B grade. Few students enjoy school if all they make are C, D, and F grades. Unhappy students dislike their teachers and don't get along well with talented or higher-achieving classmates.

To help you to understand how well we learn to make unhappy choices, I'd like to start this chapter with Joan's second focus session. You may wonder why I'm not using more people than Joan to explain the cause of unhappiness. I don't need to. As I'll explain, the source of our unhappiness, making unhappy choices as we relate to the important people in our lives, is the same for everyone. Few of us have learned to make enough happy choices to counteract all we've learned about making unhappy ones.

The Second Choice Theory Focus Session

When I opened the door to let Joan into the office, I was pleased. She was with a man who stood up as soon as he saw me, grabbed my hand, shook it, and introduced himself as Barry, her husband. He was very friendly; in fact, he and Joan seemed to be getting along a lot better than she'd led me to believe was possible.

As soon as we got into the office, Barry began to explain his being there by saying, "I've never been a fan of going to psychiatrists. I've been fighting with her to get Josh, our twelve-year-old, off that Ritalin since he went on it. Then, when she went to a shrink for our marriage, I was really upset. Who knows what he'd prescribe for her. But then she told me about you and how you're writing a book and teaching her some things that make a lot of sense to her. I got interested so she gave me the chapters you'd sent her. I liked your ideas about unhappiness. But when she was diagnosed as mentally ill, I couldn't buy into that. I agree with you. There's nothing wrong with her brain or mine. We just need to make better choices. Anyway we've been making some and already things are better between us than they've been in quite a while."

When he said that, I looked at Joan and she was smiling. They were both waiting for me to start so I began by saying, "What I want to teach you today is why you've been making so many bad choices in your marriage. There's a reason for it. Couples may deal with the reason differently but, if a thousand marriages go bad, nine hundred–plus suffer for the same reason. Remember in Chapter 4, I said that you can only control your own behavior. [They both nodded.] Have you ever seen a marriage, your parents, your grandparents, your brother's, sister's, any of your friends', or even in a movie or a book, in which after a happy beginning, one or both don't try to control the other?"

There was a longer pause than I expected and finally Barry said, "Joan's parents get along real well. Better than any couple I've ever met."

"Do they talk with each other? I mean really talk, not just a few words once in a while."

Joan said, Barry agreeing, "They do, they've always talked and enjoyed talking to each other. That's why I wanted so much to get married. Barry and I talked and talked for the first few years but then we tapered off. We still talked but we didn't enjoy it anymore. That's what you're driving at, isn't it? Every husband and wife talk, but most of the time it's not much fun."

Barry said, "She's right. Something was going on between us that made it hard for us to enjoy each other. But look at the bright side, in the last ten days we've started to enjoy talking again. We've talked more than we have in years."

Joan said, "Whatever was going on that stopped us seems to have gone away."

"If you knew what it was, you might be able to stop it from happening again. Do you have any idea what it could be?"

Joan said, "It must have to do with bad choices. You know the things we were doing, the blaming and the criticizing. We haven't been doing that since we both learned we chose to do it."

Barry could hardly wait to say something. "It's a no-brainer; I stopped talking to her because all I heard from her were complaints and criticism. Nothing I did pleased her. But what I can't figure out is where did we go wrong? We were making good choices for six years, then it was over. We're talking now but I have a feeling it'd be easy to slip back to what we were doing before."

I said, "Too easy, unfortunately. It's important for you to find out what it is. There's nothing new about it. What happened to you happens to almost everyone who gets married. Joan's parents are way off the norm."

As soon as I mentioned that we had to find out why they'd stopped talking, Joan said, "Barry, I criticized and complained because you stopped listening to me."

"Did you both stop listening because you knew ahead of time what each other was going to say? Wasn't it the same thing over and over? If you tape-recorded an argument, couldn't that same tape have told the story of a hundred spats?"

Joan said, "A thousand."

Barry added, "That's a great idea. One tape fits all. But there'd still be a problem."

Joan was a little puzzled, "What problem?"

"What would we do with all the time we'd save? At least while we fought, we listened to each other."

Joan added, "That's true, we did. . . . Boy that's really something to be proud of."

"Doc, she has point. If there was a market for marriage, we couldn't give ours away."

"If it's okay with you, I'm only going to pay attention to the place you're in now: you're talking to each other and things are better. Let's go back to the word control we talked about it in the last chapter. Can you add control to the idea you can only choose your own behavior?"

Joan said, "No problem, if I had any control over him we'd never have stopped talking. In the last two years, I couldn't get him to listen to a thing I said without starting an argument."

Barry said, "When I think of control, I think of the control freak Josh has for a teacher. When Josh stopped paying attention, she had him drugged."

Joan said, "Maybe we should put Barry on Ritalin. He has a serious attention deficit where I'm concerned."

To get them back on track, even though their comments seemed good-natured, I said, "Have you cut down on trying to control each other in the last ten days?"

Joan said, "It's not so much I stopped trying to control him; it's that he's been so much easier to get along with. It's more than listening; it's like I'm back in his life. Also, saying, 'I'm choosing this,' before I say anything to him has stopped a lot of nasty things from coming out of my mouth. He's been doing the same thing before he talks to me."

"She's right. I've stopped putting her down. I've been stopping it at work, too. People have noticed and they like it. They look at me, I guess, for an explanation. But I don't know how to explain what I'm doing without sounding strange, so I don't say anything. Anyway, what's the difference? We're getting along better with each other. I hope it lasts."

Joan said, "I hope it lasts, too, but let me get this straight. If I'm supposed to get along with Barry from now on, I have to accept everything he says and does?"

I said, "Not everything but maybe a lot more than you've been accepting for a long time."

Barry said in a very pleasant, not-looking-for-trouble tone of voice, "It's not just for you, darling. It's for me to accept you, too."

"What I'm trying to explain is, if you don't try to control each other, the problem of not talking won't be an issue."

"But what if she starts spending more than we can afford? I'm not supposed to say anything? When I say anything now, she gets angry and stops talking to me."

"I don't get angry; I start angering. How about that?"

"Okay, angry, angering, whatever it is; she gets real busy with the plastic."

"If I have to accept Scrooge's ideas about money, you better get me a lobotomy."

I said, "Joan, your mother and father don't argue; have they been lobotomized?"

"They're different; they don't pay any attention to the basic rules of marriage."

Barry added, "They are different. I love going over to their place. They never pick on each other. It feels good just being with them."

"But they have to have disagreements. How do they deal with them?"

Barry thought for a moment and then said, "I've never

really heard them have more than a minor disagreement. I've asked Joan about it and she can't remember any, either. It's spooky how they get along."

Joan said, "When I was a teenager, I saw so much unhappiness in my friends' homes, you know, divorces and remarriages, especially, the dads pulling out for younger women. My friends all liked to hang out at our house. They even remarked about my parents. One night a couple of my friends were eating with us when one of them looked at my mom and dad, and asked, 'Are you really married?'"

"What did they say?"

"I know they were kidding, but my dad said, 'No, we never got married; the married couples we saw were too unhappy.' My mom said, 'We were having so much fun dating, we figured, why rock the boat?' Then my mom gave my dad a wink and the conversation ended. Whatever they have going for them, they don't talk about it."

"But they must know that you and Barry have been unhappy. Haven't they tried to help?"

"Once I asked my mom and she said, 'Do what we did. When we first got married we had a little fight but right away we both decided it'd be our last. We still treat each other as if we're going together and it works fine.' You're the expert. How do you think they do it?"

"I think I know how. Try this. Suppose you were having an argument right now about what seems to be griping Barry, you know that Josh is on Ritalin. . . . Barry, I'm sure you have a word or two about this situation."

"You're doggone right I do. I was lucky when I went to school. When I had a boring teacher, I didn't pay any more attention than Josh does but they didn't diagnose me as having a deficit and needing a drug. It's like you said in the last chapter I read; the kid's unhappy in school and he won't sit still and keep his mouth shut. He's no more got a problem with his brain than I had."

Joan said, "But he's doing better on Ritalin. He's calmer and he doesn't talk back."

"He's calmer because he's drugged. Last year he had a great teacher and he was fine. This year I've kept my mouth shut because I don't want to confuse him. He's confused enough."

"See, Doctor, see what I have to put up with. I keep telling him . . . okay, I'm sorry. Maybe he's right. What should I do? If I take him off the Ritalin and he starts in again with the teacher, she'll flunk him. I'm the one who has to go to school when he acts up. Barry lets me have it at home about the Ritalin but that's all he does. Any time there's a problem with Josh, it's all my fault and I have to deal with it. Tell the doctor you don't do that, go ahead, tell him. Maybe you can get some help. You're here to learn something. If you're not, what are you here for?"

"Perfect. You're both doing exactly what you've learned to do all your life, criticize, blame, complain, and get angry. Joan didn't learn it at home but she saw enough of it in school and everywhere else. Tell me, are arguments like this helping your marriage?"

Joan said, "They're killing our marriage. We both know it."

Barry said, "She's a good mother. She loves Josh. Why do I start in with her like this? I'm sorry, honey, I really am. I just don't seem to know what else to do."

I said, "That's right. You don't know what else to do. Both of you don't. But you know very well what you're doing. You've been doing it for years and you're good at it. If you know how to do anything, it's how to be unhappy. Isn't it time to learn how to be happy? Or would you rather wait a few more years?"

Joan said, "Tell me you believe we can learn to be happy?"

"Why not? You're capable of learning. You've certainly

learned how to be unhappy. And you both know it's possible to learn to be happy because you've been happier since you read the material and Joan and I talked. It's not easy but Joan's parents have done it. You've even begun to do it but you're not clear on how to keep doing it. In that little tiff you just had, you went right back to square one."

Barry said, "If we want to learn how to be happy with each other, don't we need marriage counseling?"

"You may but I've been through that with Joan. If you're willing to learn what I'm going to teach you, I don't think so. I teach mental health and happiness; as far as I'm concerned they're the same thing. I used the term mental health in the beginning of the book because it sounds like a bigger deal. If I said, I teach happiness, people wouldn't listen. They'd quickly think, all I need is for everyone to treat me the way I want to be treated. You've both been waiting for that to happen for the last ten years. You can keep waiting or you can learn something new. You and Barry chose to stop controlling last week. You can choose to keep doing that. That little set-to you just demonstrated didn't do you any serious harm. You could have both chosen to behave differently as soon as Barry brought up the subject of Ritalin. You didn't have to choose to argue."

Joan said and Barry agreed, "Okay, we give up. Why do we keep doing what we just did?"

"I answer that in the title of this new chapter. I've begun to write it. It's called, *We have learned how to destroy our own happiness.* I claim in the first few paragraphs I've written that we start to learn how to make unhappy choices that destroy our relationships from the time we're very small. We continue doing it because we know a lot more about how to be unhappy than how to be happy. When you stopped enjoying each other, you knew just how to take it further. What you didn't know and still don't know is why you've learned so much about unhappy choices and so little about happy ones. Mental health is hard to come by. When you're unhappy it's easier to refuse to make love than it is to make it."

Barry said, "We're here, we want to learn. You're the teacher, teach us."

Joan said, "There's something very important here that we don't know, isn't there?"

"There is but it's more than I want to start in with now. It'll be in the next chapter that'll be finished in a week. I'll send it to you. Study it, talk with each other about it, and then we can talk about it together. But before you go, I'd like to ask you one more question. There's one group of people you almost always get along well with and you never criticize, blame, or complain about them. Do you know who they are?"

Barry said, "It's not a group but Joan's mother and father fit that description."

Joan said, "He's right about that."

Barry said, "To me, they're more like good friends than a married couple."

"I think Barry got it when he said, good friends. The people I'm referring to are your long-term, very good friends. You never criticize or blame them. You accept them just as they are."

Joan said, "I never thought about it but you're right."

"So, here's my question for you to think about. Why do you treat them better than almost all the other people you know? Try to come up with an answer before you receive what I'm writing now. It'll be in the next chapter. What we've taped today is Chapter Five, I'll send it to you along with Chapter Six."

six

Introducing External Control Psychology and Choice Theory

When we have difficulty getting along with other people, which typically occurs in our relationships as husbands and wives, parents and children, students and teachers, managers and workers, we will almost always choose to employ what I call *external control psychology,* or simply *external control.* It is a coercing, controlling, relationship-destroying psychology that, essentially, everyone in the world, no matter their culture, religion, politics, race, sex, or economic class, uses *when they are having difficulty getting along with someone else.*

When Joan and Barry began to have difficulty with each other, they immediately turned to external control and it has harmed their relationship severely. If one or both continue to use it, it will destroy their marriage.

I have given this psychology the descriptive name *external control,* because the user is always outside of the person he is trying to control. The use of external control always translates into: *I'm right and you're wrong.* The psychiatric establishment is deeply involved in external control when they diagnose you as mentally ill and pressure you to take drugs.

Based on this psychology, if you and I are having difficulty, I will attempt to pressure you into behaving the way I want you to. That can range from mild intimidation to killing you. Since all humans resist control, barring extreme poverty and severe physical illness, the use of this psychology is the major cause of human misery.

For example, we are having difficulty getting along. If I use it on you, and succeed in controlling you, you will suffer. If you use it on me, and control me, I will suffer. If, as in marriage, we often use it on each other, even if we don't succeed in controlling each other, we will both suffer. Since we choose to use this psychology just as we choose everything else we do, it is accurate to say that we choose most of the unhappiness we suffer.

We live in an external control world. It dominates the front pages of our newspapers; it's at the heart of television news, and it's essential to the plots of books, plays, movies, and operas. Without external control there's no story. We can't help learning it and we begin to use it ourselves from the time we are small as when we choose to tantrum to control our parents.

Joan's parents, who don't use it on each other or on Joan, are a rarity. But she had plenty of opportunities to learn it as she grew up by interacting with or observing the people she knew, especially her teachers and the parents of her friends. External control is by far the greatest obstacle to mental health all over the world.

In most instances, because we all believe we are right when we use it, we have no idea that we are choosing to destroy our relationships. Thousands of couples divorce every day of the year and never understand what went wrong with their marriage. Even if we become aware of the harm we are choosing to do to our relationships, we may continue because we don't know what else to do. Later in this chapter, I will begin to explain the mental health alternative to external control, *choice theory,* the psychology on which this book is based. Although I didn't cite it, specifically, everything I have written so far is directly or indirectly based on getting rid of external control.

In practice, what people do to keep their relationships from being completely destroyed is settle for the unhappy middle ground of the mental health continuum I discussed in Chapter 2. It is here that all the psychological symptoms described in the *DSM-IV* exist. But there is much more evidence of unhappiness than the psychological symptoms described in the *DSM-IV.*

There are aches and pains, such as migraine headaches, fibromyalgia, and other painful conditions for which no pathology can be found. There are physical illnesses such as heart disease, adult asthma, and eczema, which are known to be in some part associated with unhappiness. Then there are the mysterious autoimmune diseases like rheumatoid arthritis, which may also be associated with unhappiness. Earlier I referred to a small group of young adults who were suffering from this disease. What we did was teach them to replace the external control they were using with choice theory and it definitely helped.

Since we live in an external control world and firmly believe that the use of this psychology is right, to suggest that we can get along without it or with much less of it, as I will do in this book, is inconceivable to many people. It is equally inconceivable that in giving up external control, parents not only get along better with their teenagers, but paradoxically they actually gain the control they want.[1]

From this chapter on, as I already have begun to do with Joan

---

[1]My most recent book prior to this one, titled *Unhappy Teenagers: A Way for Parents and Teachers to Reach Them* (New York: HarperCollins, 2002), deals directly with giving up control in order to get more. See Chapter 14 for information on this book and others I have written.

and Barry, I will recommend that we replace the external control we are using now in our relationships with choice theory. If you do that and are successful as parents, you might want to approach your children's schools and encourage teachers to do the same. If we can get it started there, then the students, who are now the innocent victims of external control, could benefit in every way.[2]

The Specifics of External Control

External control is not complicated; no psychology used by almost everyone in the world could possibly be complicated. It is best understood by explaining the three false common sense beliefs on which it is based.

The First False Belief

The first false belief is very simple. Common sense tells us that we stop at a red light because it turns red. But in this instance, and many like it, common sense is wrong. We don't stop because the light is red, plenty of people run red lights. Choice theory explains we stop at a red light for the same reason we do everything else: *we choose to stop.* There is nothing in a red light that *makes* us do anything. If we don't want to stop, even though we may be aware we're risking our lives, we can go through.

Just because in our external control society we hear people say, "He made me do it," or "It made me do it," that doesn't mean he or it caused what we did. If we are willing to suffer the consequences of our actions, horrible as they may be at times, no one can make us do anything. It may be far from common sense but we and only we are in control of our own behavior. Accepting that

---

[2]Again, you might want to refer to my 2000 book, *Every Student Can Succeed.*

we are in control and not making excuses or blaming others or other things for our actions, is an important principle of choice theory and a vital component of mental health. This is why I asked Joan in Chapter 4 to say to herself, before speaking or acting, "I'm choosing to do this." It's a lot harder to deny responsibility when you admit you chose to say or do something.

I realize that nothing changes in your life if you continue to believe you stop at a red light because it's red instead of because you chose to stop. You're safe either way. But it makes a big difference in your life if you believe your behavior can be motivated by something or someone outside of you. A person who goes around saying *this person made me happy* and *that person upset me* is denying the fact that he is responsible for his own behavior. Men who abuse their wives, and then say that their wives made them do it, are a common example of this denial.

Choice theory teaches that all anyone or anything outside of us can do is give us information. Even bells, buzzers, lights, and whistles are information. It's your choice to heed them or not. You may use any information you have to decide what you will choose to do, or look for more information to help you make a better decision, but, in the end, whatever you choose originates in your brain and you are responsible for it. Choice theory is the exact opposite of external control; it is a responsible internal control psychology.

The Second False Belief

The second false belief of external control is that you can control someone else. Attempting to control the other was a part of Joan and Barry's daily routine, each believing that if they tried hard enough they could gain control. Barry did it with angering, blustering, sarcasm, and denial. Joan did it with criticizing, blaming, depressing, and withdrawing. Most of the symptoms described in the *DSM-IV,* along with the aches and pains I mentioned earlier in this chapter, are an attempt to control someone else or to escape someone else's control. Barry and Joan tried for ten miserable years to control each other and didn't come close to succeeding. Whether you do it gently or harshly, use a carrot, a stick, or a symptom, it is still external control.

The Third False Belief

As harmful as the previous two beliefs are to your mental health, neither is as harmful as the third false belief. This premise goes beyond harm; it goes on to destroy most of our relationships. It is the one to be aware of and to make a strong effort to avoid. It goes this way: When we are small, we recognize the advantage of getting along well with other people, so we learn to be courteous, cooperative, and to listen to what they have to say. It's useful for the rest of our lives to behave this way. It'll help us to get close and keep close to the people we need. This much is very good; we should stop here, but too many of us don't.

This is because a little later in life we slowly and inexorably get another insight: *Not only do I know what's right for me, I know what's right for everybody.* As soon as we put that insight into practice, we sow the seeds of unhappiness in every relationship in our lives. Armed with the belief, *I know what's right for you and everybody else,* we increase our use of external control. This premise is what motivates all controllers; whether they employ a carrot or stick, or force a brain drug or an electric shock on you, they all justify it with the belief, *It's the right thing to do.*

It is easy to see the third premise in despots, but it's much more often used by people who believe they are using it to do good, as the psychiatric establishment does. Politicians who fight against improving health care for the poor or support schools that fail children all justify their actions on with "It's the right thing to do." People who believe they're right are the cause of all the misery, mayhem, and murder that dominates human history. The third premise was at the controls of the planes that destroyed the Twin Towers on September 11, 2001.

But forget about politics and history. You encounter the third premise everywhere you look. Almost everyone who rears or works with children believes in punishing them when they transgress, claiming they have to be taught that there are consequences for their actions. Punishment may work; the child will bend to the punisher's will, and if the punishment is mild, logical, and just, it may not seem to do much harm.

But the risk of harm to the punisher-child relationship is always there. Who is to say what's mild, logical, and just? It is always the punisher, never the punished, who makes this judgment. A mentally healthy parent or child care worker, works to improve the relationship with the child. Good relationships, not control, are the core of mental health.

The same goes for failing children in school. Everyone knows that school failure harms the child, the school, the family, and the community. But no one in power seriously questions the validity of school failure even though through denying education, it fills our prisons and costs us billions of dollars that could be used to help, not harm, our schools.

The Seven Deadly Habits of External Control

To drive home the truth about the great harm external control does in all our lives, let me make it quite explicit by introducing you to what I call the *seven deadly habits of external control.* You've already met some of them in Chapters 4 and 5, but here I'd like you to meet the rest. Actually, once you read about these, you'll realize there are more, but if you can eliminate these seven from your relationships you will be well on your way to better mental health.

The first habit is *criticizing*. This I believe is the most deadly. In a sense, the other six are variations of this one. I can't conceive of someone who's being criticized saying, "Thank you, I needed that. I appreciate your spending the effort to correct me." People don't like to be criticized and the more accurate the criticism is, the less they like both you and what you said. Even mild criticism can do a lot of harm. One prolonged exposure may finish the relationship for good. Constructive criticism is the most moronic of all the oxymorons.

The second habit is *blaming*. When you fix blame you harm your relationship. It's not quite as destructive as criticism, but it's close. The third is *complaining*. If you're the one who's being complained about it's very deadly. But it's deadly even if you're not the target: you tend to suspect, if your turn hasn't yet come, it soon will. Even if the complainer never complains about you, it becomes an irritant to the relationship that may eventually do a lot of damage.

The fourth habit is *nagging*. This one is obvious, but there is one thing I can add that may help you if you feel you have to nag. If you've told a person half a dozen times to do something, take my word for it, he's heard you. Telling him another twenty times will only increase his resolve to pay no attention to what you want. The fifth is *threatening* and the sixth is *punishing*; they speak for themselves. Ineffective parenting and teaching are replete with these two habits.

The final habit is *rewarding to control*. In the law, it's called bribing. People like to be rewarded, but they don't like the person who rewards them if, to get the reward, they have to accept control. Don't get me wrong. I believe in rewards, but not if something is wanted in return. The quids never like the quos, and vice versa. Neither thinks he ever got enough. It's not that I don't want recognition; almost all of us do if what we've done is praiseworthy. But if it's to control me, I'm not interested.

The key to giving a reward is to decide if it will help or harm the relationship. A good message is, *I reward your effort but I want nothing in return.* Even though it's not asked for, when you are rewarded without control, you often get the idea, *I'd like to do something for the person who gave me the reward.* Accept any noncontrolling reward with thanks and, if you feel like reciprocating, more power to you.

The World Is Filled with People Who Have No Happy Relationships

While it may seem almost impossible for many readers of this book to conceive of being in this situation, there are huge numbers of people without even one satisfying relationship. These unhappy people are either the users or the victims of external control, often a combination of both. You read about the harm they do every day in the newspapers, hear about them on the radio, or see them on television. In April 2002 a nineteen-year-old expelled student in Germany killed seventeen people in his school. But those you hear about are only the tip of the iceberg. There are thousands more who never make the news, especially the millions who live lives of misery or commit suicide.

Huge numbers of the unhappy people whose symptoms are described in the *DSM-IV* lack this minimum for mental health. Melvin Udall, the male-lead character, portrayed by Jack Nicholson, in the movie *As Good As It Gets,* was a classic example of such a person. Rent the video if you haven't seen the movie. There are many more like Melvin. Probably half or more of the millions of people who populate our jails and prisons satisfy this criterion. Add to these all the addicts, the millions of alcoholics, as well as all those who beat their wives and abuse their children. None of this group is even close to mental health.

Choice Theory, the Mentally Healthy Alternative to External Control

I think you can already see the paradox with getting people to give up external control: you can't use it to persuade its users to give it up. You run into the old "Do as I say, not as I do," contradiction. You can't impose freedom or democracy on another country. All you can do is what I'm trying to do in this book: explain the harm external control does to our relationships and ask others to consider the mentally healthy alternative to it, *choice theory*.

Choice theory, the psychology of mental health, is best learned by contrasting it with external control. For example, the choice theory alternatives to the seven deadly habits are *supporting, encouraging, listening, accepting, trusting, respecting* and *negotiating differences*. There are more, but if you can begin to replace the deadly habits with these, you will quickly see a marked improvement in your mental health. The more these are used with people you are having difficulty with, the more mental health is injected into the relationship. Choice theory brings people together, external control drives people apart.

In the Choice Theory Focus Groups I describe later, you'll see how the group tries to avoid external control and to use choice theory with each other. Without an alternative such as the choice theory I offer in this book, there is little chance for the huge numbers of people in the middle of the mental health continuum to make the move toward mental health and happiness.

From the beginning, the use of choice theory can have a powerful impact on relationships. In a contentious marriage, for example, if there is still some love remaining, even if only one partner starts to use choice theory, the other will be hard pressed to continue using external control. The next time you are in an argument, argue pleasantly for a while and then tell the other person, "I see your side and you're right." Just giving in that much is almost magical in how quickly good feelings replace the disagreement. There's a lot of good and no harm in accepting that you are not right all the time.

Even in a long-lasting disagreement, if the weaker side stops using external control and switches to choice theory, as did Martin Luther King Jr. and Mahatma Gandhi, the choice theory can prevail. However, this works only if the belligerent side is losing strength and the choice theory side is willing to absorb a lot of punishment, and in the external control world we live in, this doesn't happen often. But something close to it happened after World War II, when the victors said to the losers, the war is over, let's get together and create a prosperous peace.

When people try to get external control out of their lives, they complain about how hard it is to unlearn a lesson of a lifetime. What they don't realize until I point it out is that there is one very enjoyable relationship in which we almost never use it. If we do, we usually apologize. As I said, previously, we don't use external control with our very good long-term friends.

When I explain this avoidance of external control with your friends to small groups, as I have done countless times, I ask them, why don't you use it with your good friends when you use it with so many others. They think for a moment, tend to laugh a little, then one of them usually says, "If I used it with a good friend, I'd lose her."

What makes a friend different from a marriage partner, a relative, or even from a teacher or an associate at work, is that a friend is close because she wants to be close, not because there's any obligation to be close. The friendship itself becomes the obligation. You can count on a friend to accept you as you are, never turn her back on you, and stick with you through thick and thin for a lifetime.

Almost all the other people you are connected with have

accepted that an obligation comes with this connection. There's nothing wrong with being obligated to each other, but when you are, the idea of you having to do this or that because of the obligation is always lurking close to the surface. That obligation quickly turns into resentment, then external control. Unlike a friend who wants no control over you, a spouse, a boss, a child, even a teacher or a coworker almost always wants some control over you. Also, even with a good reason, you can rarely walk away from these people.

But with a close friend, a good reason is sufficient and acceptable. You can walk away for a while, sometimes be separated for years and be welcomed back as if you were never away. For mental health, we need at least one satisfying nonobligatory relationship and a friend can be that one. More than one is better but the minimum is one.

External control is so widespread for good reason. It not only causes almost all relationship problems but it is also used to attempt to solve the problems it causes. If you remember, for years, when Joan and Barry were having problems, both thought *I'm right and you're wrong.* In that situation, there is no compromise, no attempt at negotiation.

Both thought the only way to solve the problem was to make the other give in. If he or she won't give in, increase the pressure, with the result: increase the unhappiness. It's like starting a fire with gasoline and then continuing to pour gasoline on it to put it out. In a marriage, even divorce may not put out the fire. It just allows each party to step back and avoid getting burned too badly while the fight over money and custody of the children rages on. The unhappy couples who don't get divorced stop talking to each other, which is their way of stepping back even though the fire is still blazing around them.

For me, choice theory is mental health or the psychology of good relationships. I use it every day with the people in my life.

I teach it in all my lectures and I am confident that, by the end of this chapter, you will already know enough to start using it to replace some of the external control you are using now.

While no one likes to be controlled, almost everyone likes to feel the power that comes with control. It feels so good that it's hard for you to realize how dangerous power is to any relationship you use it with. Many a parent has lost a child through too much coercive control. As soon as an overcontrolled child gets out from under your thumb, he tends to take risks the less controlled child may not take. It is a choice theory axiom that when you are dealing with young people, the less control you use the more control you get.

But before I go on to choice theory in more detail in the following chapters, I'd like to introduce you to a little of it here by answering the question that seems to be on everyone's mind when I explain external control. People ask, "I can see how harmful external control is. But what do I do, if I see someone doing something I believe will hurt me or someone close to me? Or see someone hurting anyone or breaking a law? Should I just let it go?"

Choice theory has no answer for what to do when you see a stranger doing something harmful. Since you can only control your own behavior, do the best you can with whatever you believe will keep you safe. It's when there is personal involvement that you are puzzled, such as when someone close borrows money and makes no effort to pay it back.

I'm often in a position where I am asked to offer an answer to that question and, in that situation I often say, *let it go.* I say this because from personal experience, I have found out that in trying to deal with whatever a person is doing, I may do more harm to my relationship with him than whatever harm I'll suffer by saying or doing nothing.

This can come up when money is lent to a family member or a good friend. Or when something is lent that comes back damaged or broken. Or sometimes when someone said he'd meet me and then calls and says he forgot. In these instances, I have a rule of thumb that I apply. When someone forgets an appointment, I say I missed seeing you and I let it go. Whenever I lend anything, I say to myself, don't expect to get it back at all or back in good condition. If I really don't want to lend it, I don't; for example, I rarely lend my car or more money than I can afford to lose.

I say this because I know from experience that things sometimes don't come back or come back damaged. If I want it back or back in perfect condition, I don't lend it. I realize that by not lending it, the relationship between us may be harmed. So if it's not too much money or it's replaceable, I lend it. Not lending it would harm the relationship more. If I get it back or get it back in perfect condition, I'm pleasantly surprised. To me the relationship takes precedence over things and moderate amounts of money. You can replace or repair most things. You may not be able to replace a family member or a friend.

But now let me go ahead and explain what you may do or say when you are having a problem with someone close to you when you neither want to ignore the situation nor to use the deadly habits. Let's say that you believe, as some men do, that your wife is spending too much time taking care of her elderly parents. You have had very bad experiences with her when you used the habits, but you are fed up with what she is doing. Marriage is fragile; any setback you can avoid, you should. Let me dialogue what I suggest.

Since the problem is eating at you, you have to say something, but what you say, how you say it, and the tone of voice you use are all crucial. You are treading on a marital minefield; try not to set it off. Start by saying to your wife at a time when you and she are on good terms and have some privacy, "Honey, I'm getting concerned about all the time you spend with your parents. I think we ought to talk about it."

As soon as you say that, storm clouds start darkening the

room, so be careful but continue by saying something such as, "I've been thinking about all the time you're away from the house, especially on weekends."

She'll still be defensive, thinking you're more able to take care of yourself than her parents are. She's already pretty sure that complaining and criticizing are about to start. Here is where I suggest that you come through with the last thing she'll expect in this situation: a choice theory inspiration that will bring you closer to each other than you are now. But by now she'll feel she has to say something, so she says in a somewhat defensive tone of voice, "It's hard enough doing what I have to do without getting static from you."

But you say, "Instead of me sitting home alone, I want to go with you on some weekends and I want to stop by myself once during the week. I've been sulking like a child and it's ridiculous."

All of a sudden the storm clouds lift and the room is filled with sunshine. Very likely she hasn't had this much support from you in a while and she'll appreciate it. This is what choice theory is all about. You have everything to gain and nothing to lose except some time by yourself, which you don't enjoy anyway.

When you begin to understand choice theory, you look for every chance you can get to say or do something that brings you closer to the people in your life. You are in an intimate relationship; you know what she wants. Offering help and companionship before she has to ask is what keeps love alive. Sitting around alone stewing and hoping things will get better is the least effective thing you can do.

For choice theory to work, you need to keep it up. If you slip, apologize immediately. Choice theory is fragile in the beginning, but once you grasp it, it's very sturdy. External control is in the saddle now in most marriages. It seems strong because it has no opposition. But if you'll give choice theory a chance, external control will melt away like a late spring snow.

seven

The Third Choice Theory Focus Group Session— Joan, Barry, and Roger

Joan and Barry were with a third man who looked to be in his mid-sixties. Joan introduced him as her father, Roger, who said, "I hope I'm not intruding but I've been reading all the material you've been sharing with them and I asked if I could come along."

Joan said, "He's known about your work for a long time. He read *Reality Therapy* years ago."

Barry said, "We've talked to you about him. Roger lives what you're trying to teach us. We figured that if you got so busy you couldn't meet with us, he could take over."

I said, "Roger, glad to have you. Have you read any of my recent books?"

"No, to tell you the truth, all I'd read before I found out you've been working with Joan is *Reality Therapy*. I bought it while I was in my training to be a psychologist and I've been using it ever since. But I did get a copy of *Choice Theory* a few weeks ago. Both Jean (she's Joan's mother) and I've been reading it. She'd have come with me but I thought two from our family would be enough. I guess you could say we've been choice theory people almost all our lives. If there's anything we can do to help you, Jean and I would like to get involved in this project."

"Just read, listen, and discuss. This project is open to anyone who wants to improve their mental health or help others to improve theirs. Can you buy into the idea that there's no mental illness unless you've got pathology in your brain?"

"I've read several of Peter Breggin's books,[1] and I agree with him. Nothing you said in Chapter 2 surprised me."

Joan said, "He loves the idea of teaching mental health. That's why he's here."

"Roger, you're more than welcome. We need all the help we can get. How do you think Joan and Barry are doing with what I sent them since they were here last?"

Roger said to Joan and Barry, "Go ahead, tell Bill what you've been telling Jean and me."

Barry said, "We really liked what you wrote about external control."

Joan added, "Half the world's trying to control the other half. We really do live on a planet dedicated to external control. But how did it get that way. You didn't explain that."

"Your dad knows how. It's in Chapter 2 of *Choice Theory.* Did he explain it to you?"

Roger said, "I didn't explain it because I felt you had a reason for holding back. This is your show, Bill, not mine."

"It's in the next chapter, Joan. I think you'll find it interesting. It's already written. I can give you a copy when you leave today."

---

[1]Peter Breggin, M.D., is a psychiatrist I've been associated with for several years. See the Appendix for some of his salient work.

Barry said, "External control has the whole world screwed up. For the first time, I really understood my parents and my grandparents. The deadly habits were alive and well in my family. What I can't figure out is how Roger and Jean escaped them."

Joan said, "Don't worry about Roger and Jean. We have to make sure we stop teaching them to Josh. Since he got into trouble in school, we've been "deadly habiting" the poor kid right and left. No wonder he likes to spend time at my mom and dad's house."

Roger said, "Josh'll be okay. Explain what you've been doing, back off, and he'll be fine."

I said, "That's what I was trying to explain. Things get better real quick when you stop using the habits."

Barry said, "It's almost as if they reinforce each other when you use them. I manage a bunch of salesmen. The whole atmosphere's changed since I stopped using the habits. All I did before was criticize, blame and complain, nag, and threaten. We'd even bribe to try to get them working. But it all backfired. All we did was argue, put each other down. Now I tell them that my job is to do what I can to help them. I'm through with the habits. I explained the whole thing to them. It's amazing how quickly they caught on. Once in a while when somebody slips, we point it out, laugh, and it's over."

I asked, "How'd you do on the question I asked you? Why do we get along better with our friends than with anyone else?"

Joan said, "We didn't get it. As soon as I read your answer about obligatory relationships like marriage I caught on. I remembered my friend Janice who asked if mom and dad were really married. She saw they loved each other without being obligated to each other. We tend to see obligation, even love, as external control, don't we?"

Barry chimed in, "I hated it when the guy that married us

said, love, honor, and obey. I don't mean the love and honor part, I mean the obey part. I've noticed they've mostly cut that out of the ceremony. That's a step toward mental health, isn't it?"

Roger said, "A judge friend of mine who marries people has stopped saying 'till death do ye part.' He's been the judge in so many divorces that he now says 'as long as it lasts.'"

Joan said, "If it weren't for Josh I think we'd have gotten a divorce. But I'll tell you, being a mother is a lot harder now that I'm giving up knowing what's right for him. Let me tell you what I'm worried about. The high school he's going to in a few years is full of druggies. I've heard him talking to his friends about it. I also hear him making fun of the drug prevention program they use in his school. It's full of *we know what's right for you.* But what else can you do to try to keep kids off drugs? I'll admit I know what's right for Josh where drugs are concerned. How do I avoid using external control if I find out he's experimenting with drugs or alcohol? What's the choice theory approach to that problem?"

I said, "Roger, you're his grandfather. What would you do if Josh told you he was smoking dope?"

Roger said, "Joan, what did we do, your mother and me, the time you went to that party when you were fifteen and got so drunk you passed out? The kids got so scared they called us. We came right over but, thank goodness, you woke up enough when we yelled your name so we took you home instead of to a hospital. Do you remember that?"

"Oh, wow, Dad, I'd forgot about it."

"Maybe you forgot getting drunk; a lot of people forget that. But you haven't forgotten what we did about it. We got you home and you slept a long time. We were there when you woke up. You remember that?"

"I remember being home in bed but I didn't remember how I got there. When I asked you, you told me that I'd gotten drunk and passed out at the party. I thought you were going to kill me for what I did. I'd been doing a lot of drinking and now you knew."

"What did we do?"

"You asked me how long I'd been drinking or was this the first time."

"What did you tell us?"

"I told you it was the first time."

"I think it was your mother, what did she say to you?"

"She said, 'We have no intention of punishing you. But if you'd tell us the truth, we'd appreciate it.' I don't know how she knew I was lying."

"Then what did you do?"

"I cried my eyes out. I just couldn't seem to stop crying. You both stayed with me, and then Mom said, 'It's okay, Joan, you can tell us the truth.' So I told you I'd been doing a lot of drinking. And I told you I'd also been smoking pot. And then I started to cry again. When I finally stopped you both said, 'We love you. We haven't stopped loving you because you've been drinking and smoking dope. But if you ever do anything at all that you know we'd be concerned about, please tell us the truth. If we don't know what's going on, we can't help you.' You then asked me if I thought I needed help or if I thought I could handle this on my own. I said I thought I could handle it on my own."

"So what did we do?"

"You told me that all you wanted is for me to tell you and Mom the truth so we could decide if you were handling it okay. I've told you the truth ever since. You know about a few other things I've done but I've never been in serious trouble."

Barry said, "Roger, I've been involved in some of the other things. I wasn't used to telling adults the truth, especially when their daughter was involved. You helped both of us. I never have been able to talk to my parents like I have with you and Jean. When you can talk to someone you love like we can to you, you don't screw up your life over the long haul."

Joan said, "We've got to talk more to Josh. I was so full of being a know-it-all teenager that I stopped talking to my mom and dad when I started to drink and have sex. To them, I was such a goody-goody, I never suspected that Mom was on to what I was doing."

Barry said, "Well, we won't have that problem with Josh. When it comes to getting into trouble, he's a lot more like me than you."

I said, "You know what I've found out. I think teenagers and even preteens like Josh need to talk to their parents more than when they're little. In the world we live in, they just assume that parents will punish first and ask questions later if they get into any trouble. What boys need, if they're lucky enough to have a father, is time alone with him that they can count on. But the time needs to be quality time, which to me is time without any external control. If I were you, Barry, I'd arrange to spend an hour or two on Saturday with Josh, if he's willing. I think that'd prevent more trouble than anything else you can do."

Roger said, "Girls need it, too. Both boys and girls seem to have enough quality time with their mothers; they need time with their fathers. I have teenagers in my practice. I've called some of their fathers and told them if they'd spend time with them they wouldn't need to see me. But their fathers were usually too busy or had another excuse. It's almost as if they're afraid to spend time alone with their own teen."

Barry said, "When Bill said that, I said to myself, What'll I do alone with Josh every Saturday for two hours? I felt a little scared."

Joan said, "I felt for Barry when Bill said that. He never had a relationship with his own father. I can see how awkward he is with Josh unless he knows just what to do."

After she said that, there was a pause and then, without saying anything, Barry got up, went over, and hugged Roger. In a moment he began to cry silently. We could all see his chest heave as Roger put his arms around him. "Barry, it's all right. I'm your father now. I can help you with this. We'll talk."

Barry, tears still in his eyes, said, "But I almost lost you. If Joan had left me, I would've lost you. I'm forty years old and I need you just like Josh needs me. I can't screw up my marriage and I can't screw up with Josh. I can't."

Joan rushed over to her father and he held both of them. She started to cry and said, "Barry, you won't lose me. You won't lose Josh. If we stop trying to control each other, we'll all be okay."

To tell you the truth, I felt like crying, too. But at the same time I felt good. "Roger, I guess you're on the team now whether you like it or not. Joan, Barry, it's enough for today. Go home and love each other."

eight

The Role of Our Genes in Our Mental Health

When Joan and Barry began having trouble in their marriage, they had no idea that their genes had something to do with the problem. As they read this chapter I think they will be even more determined to keep external control out of their marriage. Choice theory explains that we are born with purpose built into our genes. For example, if you have no food, your genes provide the pangs of hunger that motivate you to eat. You don't have to learn how to be hungry.

But what your genes do not provide is what you have to do to find food. That, you learn on your own. Essentially, our genes provide us with five basic needs that motivate all our behavior from birth to death. But unlike all other creatures, they do not provide us with the behaviors we need to satisfy these needs. Barry was motivated by one of his basic needs when he practiced the deadly habits on Joan. But the deadly habits are not genetic. They and all other human behaviors are learned. And they can be unlearned, as Barry is doing when he begins to replace criticism and blaming with supporting and encouraging.

The five basic needs are *survival, love and belonging, power, freedom,* and *fun.* All living creatures have survival built into their genes or there would be no life on earth. In humans, species survival is the need that drives much of our sexual behavior. All mammals (including humans), birds, and some reptiles have the love-and-belonging need built into their genes. If they didn't, they wouldn't take care of their young and their species would die out. The needs for freedom and for fun are needs that help us, as well as many other higher mammals, to survive. But the need for power is essentially human.

Lower mammals may fight to defend a territory in order to get enough food. The males may fight to get a mate so that the winner's genes strengthen the gene pool. But only humans fight to dominate and control other humans when neither food nor genes are at stake. It was Barry's need for power that led him to try to dominate his marriage. Power is the need that has led the whole human race to practice the external control psychology that harms our relationships, makes it almost impossible for us, over time, to satisfy our need for love and belonging in obligatory relationships , and is, by far, the main source of our unhappiness.

We have no way to remove the need for power from our genes any more than we can get rid of our survival, love, freedom, and fun needs. Children who were loved survive in larger numbers than children who were less loved. More people who are free survive and more people who have fun survive because they learn more: fun is the genetic reward for learning. And obviously, powerful people survive in larger numbers than the weak. This is how the genes for these needs became built into the human genome and supplement the basic need to survive.

But driven by our needs, especially the need for power, we have now learned enough so that the very survival of the human species hangs by a thread. If we don't succeed in restraining our need for power enough to keep nuclear weapons under control, we are gone.

If we don't learn to replace external control with choice theory in our relationships, it is doubtful that we will ever become much happier or more mentally healthy than we are. In affluent First World countries, individual survival is no longer a problem. Marriage, family, school, and the workplace, however, present huge relationship problems that will not be solved as long as almost all of us embrace external control.

Although I think it is obvious, the genetic reward for behaving in ways that satisfy our basic needs is that we feel very good. But as I have already explained, there are two different ways to feel good. The first is in loving and caring relationships, which I call *happiness*. The other is feeling good without love or friendship, which I call *pleasure*.

Even though this may already be apparent to you from what I've written, the major problem facing the human race is how we can satisfy our need for love and belonging and still satisfy our always pressing need for power. As long as we satisfy our power need with external control, we will never solve the built-in conflict between two needs: love and power. It was this conflict that suddenly and tragically confronted King Midas when he touched his daughter. As far as I can figure out, the only way to attain power that is not destructive to our relationships is through *mutual respect*. But as long as we practice external control, mutual respect will be hard to achieve.

When Barry criticized and blamed Joan, he did not respect her. When Joan depressed and withdrew from Barry, she did not respect him. With little respect, both were unhappy and very likely respected neither himself or herself. But unless huge amounts of anger surface between them, they may be able to hang on to their marriage and keep the love of their son. But hanging on is hardly the way to spend a life together.

However, the human race has a lot more to lose if we don't learn to respect each other regardless of our differences. So far we have been able to avoid nuclear disaster through mutual fear of annihilation. But fear of death is a rational concern based on simple reasoning: if I do this I may not survive. But if people are willing to die based on the belief they'll find their reward in heaven, fear is out of the picture. Again, respect for all people, whether they differ from us or not, is humanity's best chance of surviving another century.

The sad situation in the external control world we live in is that when we try to gain respect through power, as Barry did with Joan, our efforts often backfire; instead of reducing his external control, Barry increased it as the years went by, just as Joan increased her depressing and withdrawing. Of course neither knew what he or she was doing. But what happened is what almost always seems to happen in relationships: the more they tried to control each other, the less control they got and the more the relationship was harmed.

This process is epidemic in our schools. If children don't respect both us and what we are trying to teach, and refuse to pay attention, we label, punish, and drug to try to control them. We refuse to accept the evidence of our own eyes: there is no ADD or ADHD in the classes of teachers who use choice theory. There is a much better way to educate children, but so far most of the people who run our schools refuse to stop using external control.[1]

Only Family Love Is Genetic, Infatuation and Romantic Love Are Not

Since good relationships are the core of mental health, it is important to understand that only family love and belonging, the love that helps us to survive, is programmed into our

[1]See *Every Student Can Succeed,* referred to in Chapter 14.

genes. Mostly this is expressed as parent/child love but, since it is in our genes, it usually extends into the greater family.

As important as family love is, it is not the love we struggle for as adults. Sexual love or romantic love, not sex itself, is learned and can very quickly be unlearned. The tragedy of intense but short-lived romantic love, almost always ended by external control, has been the heart and soul of all literature—of novels, stories, poetry, plays, movies, and opera—from the time these forms were created. Happiness prevails when neither partner tries to control the other, as exemplified in this book by Roger and his wife, Jean. But that kind of happiness is a nonseller at the box office and a rare occurrence in life.

The lasting love we got, still get, and give to our children, parents, grandparents, siblings, close relatives, and friends is in our genes. For most of us this love lasts a lifetime, and it can and does survive a lot of external control. The television series *All in the Family* represented an excellent example of this happening. Archie Bunker was Mr. External Control himself. He loved his daughter and his wife as members of his family, but there wasn't much romance between Archie and Edith. There was, however, a lot of family love behind his control, or the show wouldn't have lasted more than a season. Unlike romantic love, which no one expects to last, the portrayal of genetic family love prevailing over external control gave the show its popularity. People could relate to it.

Many of the letters that fill the "Dear Ann" and "Dear Abby" columns are from married women, women living with a man, or women trying to get a man to marry them. They ask, What can I do to get a man to care for me the way I want. Basically, they ask for advice on how to control the man when they sense their romance is drifting away and they have no idea why. What love most of them can count on, if they can refrain from using too much external control, will be with their children.

In the external control world we live in, sustaining an enjoyable romantic, committed, sexual relationship for a lifetime, as Roger and Jean of the last chapter have, is beyond the grasp of well more than half of all married couples. A recent article in the *Living* section of the *Los Angeles Times* (September 11, 2001) describes a new church or synagogue divorce ritual in which the ex-husband and -wife celebrate, in the presence of the minister or rabbi, that they have agreed that their divorce is not the end of caring and support for each other and their children. The article goes on to say that many people believe that this ritual sends a bad signal: it encourages divorce.

I don't agree. I think what they are doing, *albeit without knowing it,* is sending each other the message that, even though we can't live with each other without practicing external control, we don't have to carry this control into our divorce because we are no longer locked together under the same roof.

This is certainly a step in the direction of mental health for the couples who participate and, especially, for their children. They obviously still want to be friendly and have no idea what really went wrong. If they had known more about the damage external control can do to love, they might have been able to preserve their marriage.

Although marriage may be the oldest of all human traditions, what genetic support there is for it may come from having children and from the greater family. The loss of love in the marriage after the children leave has been given a name, the *empty nest syndrome.* If both parents enjoy rearing the children, that happiness often spills over into the marriage itself.

I believe that there are more mentally healthy married women than there are married men. Because women are the primary caretakers of the children, they have figured out that it is to their benefit to reduce their use of external control in their marriage. With less external control, they find they can still have some spontaneous or even planned romantic moments. Even though there may be less romantic love, with less external control, they can almost always count on more support and friendship from their husband. Add to this the lifelong genetic love they and their husband get from the children and family, and they have the makings of a strong marriage. Is this fair? I don't think so. Is it wise? I think it's very wise.

Infatuation, one of life's most enjoyable but least lasting experiences, is so strong because the need for power is temporarily subordinated to the desire for romantic love and sex. I'm not saying that infatuation can't turn into love but it rarely does. As soon as the sex gets the least bit routine, external control surfaces and kills the romantic love.

Unfortunately, women, more than men, get fooled by infatuation into thinking it's love because it feels so good and the media portrays it as love. When the women find out it isn't love and get turned off, the men sometimes become violent. Ending an infatuation is a dangerous time for a woman. Infatuation, therefore, becomes for both men and women, the lifeblood of affairs. When the infatuation ends, the affair should end and usually does. Trying to keep it going is like beating a dead horse.

One of the main purposes of mental health education would be to teach teenagers, both girls and boys, about the mirage of infatuation and the dangers of romantic love. Neither men nor woman are capable of sustaining romantic love for very long. But statistics show that men are less capable than women of sustaining genetic love because they are more power driven and more prone to using external control. Teaching about the destructiveness of external control would be the core of mental health education. You can't correct what you don't know about, and external control is one of the world's best-kept secrets.

External Control and the Other Basic Needs

External control is not easy to give up or avoid as we try to satisfy any of the needs. For example, to survive you can get a job and support yourself. If you are lucky enough to work for a mentally healthy boss who doesn't use much external control, you are likely to be happy at work. Some bosses have learned about this; it is a part of many management training courses. Managing this way makes more money for the company by increasing the quality of the services and the products. If you don't want to leave your job, you can do what the characters in the Dilbert comic strip do. They're far from happy but they do have a few laughs as they struggle against their idiotic, controlling boss.

To satisfy your need for freedom is not too difficult unless you're in a very controlling marriage. But if you, yourself, are not too controlling, if there's still some love left in the marriage, and you explain to your mate what is in this book or read it with him or her, you may get the freedom you need and help your marriage at the same time. Remember, choice theory continues to say: It's what *you* do in a relationship that counts, you can't control your mate or anyone else.

Fun is also very easy to satisfy if you know what it is and don't try to get it by putting another person down. But when I tell an audience what I've written in this chapter, that fun is in their genes, some people get upset. They think I'm accusing them of having frivolous genes and, with some heat, they say, there's no fun in their genes. I always agree with them; for all I know they may be right.

What they fail to hear is that the experience we call fun is not frivolous; it is the reward for learning or teaching something useful or sharing something enlightening with someone else, often a joke. Some people enjoy learning by themselves, but it's no fun if you don't share what you've learned and maybe get a laugh or two in the process.

If you ask a child, "Who is a good teacher in your school?"—something I've done many times—invariably, the child mentions the name of one teacher. If you ask why this teacher, the child says that he or she *makes learning fun.* What that teacher is doing is tapping directly into the child's genetic need for fun, just as a parent who loves a child taps into the child's need for love.

Since fun is the reward for learning something useful, it has a lot to do with all the other needs. It's fun to get a good job and take care of yourself. It's fun to fall in love and make an effort to learn new ways and places to enjoy sex. I call sex the recreation of love. It's also fun to figure out all the ways you can show respect for each other in a relationship.

A lot of fun in a relationship makes it special and fills it with laughter to keep it special. Fun is the glue that binds the needs together and allows us to satisfy several needs with the same behavior as when we get involved in a recreational activity. Fun can be frivolous, even slapstick. Once in a while, for a few minutes, I like to watch the Three Stooges.

Listen to a good comedian like Bill Cosby for an hour and he'll keep you chuckling as he paints accurate pictures of real life. Fun is when you separate truth from hypocrisy. Both great comedians and great teachers have the knack of doing this.

Some Final Comments about Satisfying Our Needs without External Control

A saying has been attributed to Mark Twain: "Man is the only animal who eats when he's not hungry, drinks when he's not thirsty, makes love any season of the year, and learns nothing from experience." It's the last part of this quotation I want to talk about because one of the criteria of mental health is that we make an effort to learn from experience what external control does to our relationships.

Earlier I said that you're given a warning when you're about to say or do something that is likely to disconnect you from an important person in your life. I can now explain that this warning is built into your genetic structure. You are so driven by your need for love and belonging and have had so much harm done to your relationships by the use of external control that your love-and-belonging genes will recognize an impending disconnection even before you do and warn you to desist.

As soon as you think of something disconnecting and prepare to say it, you'll get a quick feeling of discomfort, maybe similar to the feeling of discomfort you get in a strange place when you hear a sudden noise when your survival genes kick in to give you warning. My advice is to pay attention when you get that warning before you say something that may harm a relationship that you don't want to harm. If the warning was wrong, no harm can be done by keeping quiet. If it's correct, as it almost always is, it will have served you well.

If you have a day when you feel very good, you'll know that you succeeded in satisfying your basic needs. If you have a bad day, it's the opposite. A mentally healthy person pays attention to feelings. I've found out that when I get even a small discomfort, I take a moment and say to myself, "Okay, which need is it?"

It's probably not survival; that would be obvious. The same goes for freedom. If you're being totally controlled, you're certainly aware of that; you don't need to look further. It could be love and belonging, most likely belonging because you're usually aware of your love situation. But frequently it's fun. And if you're in a boring, nonlearning situation, if you make an effort, you can probably do something more need-satisfying. Instead of complaining about how you feel, pay attention and have some fun.

Through public awareness that we all should have a chance for happiness, we have spent huge amounts of money to create laws, a judiciary system, schools, colleges, social services, medicine, psychiatry, psychology, armed forces, and prisons. What we don't understand is that almost all of this effort is an attempt to reduce the harm done to us by the excessive use of external control. We use it and it's used on us. Much of the need for these expensive efforts could be reduced if we would stop focusing on unhappiness and mental illness and start focusing on happiness and mental health.

nine

How Can You Say That We Choose Our Symptoms?

As soon as I finished Chapter 7, I sent it along to Joan and Barry and started to work on this one. About a week after they received Chapter 7, I got a call from Joan. As soon as I heard her voice, I was concerned that she and Barry were having trouble with the basic needs, but she said, no problem, the chapter pretty much spoke for itself.

She then went on: "About a week after we got through with Chapter 7 [in which Barry and Joan end up crying and hugging Roger], the book club we've been attending for years held their regular monthly meeting. After the meeting, everyone stayed for refreshments. We started to chat and I decided to tell them about you and the book you're writing and how Barry and I have been giving you feedback after reading the first six chapters. I brought up the idea that we choose most of the psychological symptoms we suffer from and explained I'd been choosing to depress for quite a while, but I've learned to make more effective choices and I'm feeling much better. As soon as I said that and saw the funny looks on their faces, I went ahead and explained the difference between being depressed and choosing to depress. It helped a little but I'll tell you, the idea that we choose our symptoms created quite a ruckus. A woman named Bev was adamant. She said, 'I'm depressed all the time and believe me, I'm not choosing it.' Then Jill, another woman, who usually doesn't say much, said she has migraine headaches and, if I believe she's choosing them, I'm crazy. To tell you the truth, you've been talking about pain since the first chapter and Barry and I've been having some real trouble with the idea we choose to hurt. But then it got worse. Jeff, a young guy, I guess in his late twenties, said, he has rheumatoid arthritis and hurts all the time. Do you say he's choosing that, too? C'mon, Dr. Glasser, I have an aunt with rheumatoid arthritis. You can't be saying that's a choice? I guess you can see how confused we are. Anyway, another man, Neil, said his sister Amy has panic attacks but he wasn't as skeptical as the others. She's been doing it for two years and he suspects she can turn them on and off whenever she wants to. What I want to know is, do you want us to stop talking about what we've been learning? It really stirs people up."

"Don't worry about stirring anyone up. You got a little ahead of yourself but I'm glad it happened. Believe me, I'm not trying to hide my ideas. I'm looking for people to challenge them; it helps me. Besides, there are some choice theory answers to all those questions. I've been working on them in Chapters Eight and Nine, which you haven't read yet. I'll get them to you in a few weeks. How many in your group got involved in that discussion?"

"Everyone to some extent but the four I mentioned really got into it. Oh, I hate to tell you this; there was one more, Selma. She has a son with schizophrenia and she got so angry, she said psychiatrists who don't believe in mental illness ought to have their licenses revoked. Barry and I stood our ground by explaining that the real problem was unhappiness and that when we're unhappy we can choose almost anything. But I'll tell you, the word choice is a red flag."

"Do you think those five people would like to read the material you and Barry have, plus the chapters I'm going to send you. I can easily make copies and get you as many as you need."

"I can find out. If anyone's interested, I'll call you."

"Good, but be sure to tell them there's still more to be explained than what's in those chapters. What I'm doing now in Chapter Nine may clear up a lot of their concern about choosing how they feel. Right now they're so angry they won't entertain the idea of how much better their lives could be if they understood choice theory. The only really hard question was from Jeff, the young man with rheumatoid arthritis, but there was material earlier in the book that may help him, if he ever gets a chance to read it. Tell them I'm looking for hard questions, but make it clear this isn't therapy. It's teaching mental health and the book I'm writing is the textbook."

There's still a lot to explain if people who have symptoms are going to accept my ideas. The woman called Bev really got hot under the collar. She could be having a problem satisfying her need for love. In the external control world we live in, with so many people using the deadly habits on each other, that's the most difficult need to satisfy. Someone close to her—a husband, a boyfriend, or a teenage child—is most likely the person she can't reach. External control isn't working and she doesn't know what else to do.

I believe if Joan hadn't mentioned anything to the group about choice and choosing, she'd have had more success. Maybe she could have said she and Barry were unhappy for a long time and now she felt better. Then tell the group that they're both reading a book that's in the process of being written and it's helping them. That way the group might have been interested and no one would have gone on the attack.

Everyone with symptoms believes that if they could just get rid of the symptom, they'd be happy. That's why so many people turn to drugs. It's much easier to look for happiness in a pill than to try to repair an unhappy relationship or build a new one.

Millions of people like Bev are suffering symptoms because they don't know how to get along with an important person in their life. As long as they can't get along, they are trapped in a vicious circle that closes their minds to improving the relationship. They are depressed because they're so unhappy and they're so unhappy because they're depressed. It's this trap that led Bev to say she's depressed all the time.

Bev's telling the truth; she is depressed all the time. Since you've read this far in the book, you know that to escape this trap she has to learn to replace the external control she's using in an important relationship with choice theory.

I'm banking on the fact that people like Bev can begin to learn choice theory from a book—at least learn it enough to be willing to get together with a group of people who are also trying to learn some choice theory. And find someone like Roger to help the group get started.

There are several thousand people who have been trained by the William Glasser Institute in using choice theory, and who might be enlisted to help. Most of them would enjoy helping a study group get started. But I can see people like Joan and Barry leading a group once it gets started. Nonprofessional leaders like them would emerge from every group.

Joan was trapped in the same vicious circle with Barry as Bev is with her problem person. The more he criticized her for putting Josh on Ritalin, the more unhappy she was in her marriage, and the more unhappy she was, the more she depressed.

Joan and Barry are escaping this trap with choice theory. The fact that they're more than willing to share what they've learned will make them valuable members of any Choice Theory Focus Group. Right now Bev doesn't know anything about her basic needs and external control. All she knows is that she wants the other person to change and be the person she wants him or her to be because *she's sure that what she wants is right.* The more she tries to change the other and fails, the more unhappy she is and the more she'll depress. She's trapped in the circle not because she is incapable or unloving. She's trapped, as are millions of others like her, because external control is all she knows. In the external control world we live in, she, and millions like her, are depressing; it's all around us. Depressing is by far the most common symptom we choose when we're unhappy.

By now, Joan and Barry have learned enough choice theory to put it to work in their lives. What they haven't yet learned is the additional choice theory they need to know if they are to explain why, when we are unhappy, it's correct to say we choose our symptoms. The people Joan talked with are depressing, paining, and panicking. Selma's son is suffering from what is called schizophrenia. But none of them has any idea of where their symptoms are coming from. They think what they are experiencing is happening to them, that their symptoms have nothing to do with the behavior they are presently choosing.

All of Your Behavior Is Total Behavior

I believe that all symptoms, painful, frightening, crazy, disabling, possibly even the symptoms of a disease like arthritis, are your brain's way of *warning* you that the *behaviors* you are presently choosing are not satisfying your basic needs. Your brain has evolved to help you survive. If you disregard that warning and do nothing to increase your need satisfaction, you are stuck with your unhappiness. From that unhappiness, your symptoms begin and often escalate.

By expanding the concept of behavior to the more comprehensive concept *total behavior,* choice theory can offer a way to understand that you choose your symptoms. As simple and clear as the word *behavior* seems to be, it is much more complicated than most of us realize. Up to now, when I talked about choosing behavior, I was aware that there was much more to explain than I led you to believe. I didn't think you were ready for this explanation earlier, but since Joan and Barry got into that heated discussion, I think you need it now. To begin, let me say: *All we do from birth to death is behave; all our behavior is total behavior and all our total behavior is chosen.*

The Four Components of Total Behavior

Right now, as you are choosing to read this page, you are *choosing the total behavior of reading.* It is a total behavior because it is made up of four separate components: *acting, thinking, feeling,* and *your physiology.* It takes all four components, *working together,* for you to read this page. For example, right now you are *acting* to hold the book and your eyes are moving across the lines and down the page. You are also *thinking* about what you are reading, *feeling* something as you absorb the knowledge, and finally, because you are alive, there is some *physiology* going on in your body. Certainly, your heart is beating, your lungs are exchanging carbon dioxide for oxygen, and your brain chemistry is working normally or you might not be able to understand what you're reading.

A total behavior is always named after the most obvious of

the four components. In the last paragraph you are "thinking" about what you are reading, therefore the total behavior would be called "reading." Last night you chose the total behavior of "sleeping," because that was the most obvious of the four components. This morning at breakfast, you chose to eat, or the total behavior of "eating." Perhaps to get some exercise, later today you may choose to run, that is, choose the total behavior of "running."

Once you accept that all your behaviors are total behaviors, then you'll realize that all your total behaviors are chosen. Even though you may say to someone who doesn't know choice theory, "I'm depressed," to yourself you would be thinking, *I'm depressing,* or *I'm choosing to depress.*

For example, Bev got angry when Joan and Barry told her that she was choosing to depress. For her, it didn't feel like a choice. All she felt was the *feeling* or *depressing* component. She had no idea that three other components were also involved at the same time as the feeling component.

What choice theory will teach her is, instead of thinking *I'm depressed,* to think *I'm choosing to depress,* or, even better, *I'm choosing the total behavior of depressing.* When she learns that, she will also learn that her present *actions* are to use external control on someone. Her present *thoughts* are, This person is not doing what I want her to do; her present *feelings* are the feelings of both depressing and angering, and her *physiology* is probably feeling a lot of fatigue caused by trying to control someone day after day and failing.

If she understood this, she would also understand that unless she chose to stop trying to control whoever it was, she would continue to depress. Before we understand total behavior, all we focus on is the component that is most obvious to us. We believe that component is happening to us and that we have nothing to do with what's causing what we may feel, think, act, or our physiology.

Bev can't stop the feeling part, her depressing, but she can stop using the external control that's harming her relationship with an important person in her life as Barry did with Joan. When he chose to stop controlling her, his angering melted away and Joan stopped choosing to depress. The same can happen with Jill's headaches if she can stop controlling or figure out how to escape from someone else's controlling. Jeff and Amy will be able to reduce their symptoms if they can accept that Jeff is choosing the total behavior of arthritising, Amy, the panicking, and make a better choice. In the next chapter we will see how that can occur as the book club group starts to put choice theory to work in their lives.

Creativity, the Wild Card of Our Brain

So far I have explained that unhappy total behaviors that people choose can have painful, sick, or panicking components that they can get rid of if they can choose more need-satisfying total behaviors to improve their relationships. Even crazy thinking, as I'm sure is happening in the brain of Selma's son, who has been diagnosed as schizophrenic, could be reduced or eliminated if he could satisfy his needs better.

To begin, I want to stress that Jill is not imagining her migraine headaches. There is also nothing imaginary about Bev's depressing, Selma's son's delusions, or Neil's sister's panicking. While Jeff's rheumatoid arthritis has a definite physical pathology, there is no known reason for his immune system to attack and inflame his joints. But just because there is no pathology to explain your symptoms, that does not make what you experience imaginary. It's as real as if there was pathology.

What still needs to be explained is where the depressing, sicking, paining, or crazying behavioral components come from. So far only depressing has a simple explanation. We learn to depress when we are very young from the people around us. Depressing is so common it's almost impossible not to learn it from your parents, grandparents, family, and friends.

Here I would like to offer a choice theory explanation for almost all the other symptoms that seem so inexplicable now: *They are created in our brain when we are unhappy.* But as you well know, they are hardly the only experiences our brain creates.

While it is not a requisite for creativity, unhappiness is the force that motivates the creativity inherent in our brain to come up with, not only the symptoms described in the *DSM-IV,* but also aches, pains, fatigue, and even some or most of what is seen when our immune system attacks a normal tissue in our body as in autoimmune disease. Creativity, helpful or harmful, good or bad, can be expressed in all four components of our total behavior.

Creative *actions* are seen in athletics and in dance. Anorexia and bulimia are creative activities some unhappy people create to control eating. Creative *thinking* is all around us, ranging from the theories that led to the release of nuclear energy to the delusions and hallucinations of schizophrenia. Even our nightly dreams appear to be a creative way to get the maximum benefit from our sleep. *Feelings* are expressed in the creativity of literature, drama, music, and art. Leonardo da Vinci asked us to try to figure out what the *Mona Lisa* was feeling. Creative *physiology* may be the explanation for the endurance of marathon runners or the pathology of autoimmune diseases such as rheumatoid arthritis and lupus.

All of us experience creativity continually and, at times, it is so apparent we are surprised. For example, have you ever been insulted and immediately racked your brain for a witty comeback? But nothing came. You were rankled for a while but then got busy and forgot about it. Hours later, a perfect response popped into your mind. Late as it was, it still felt good to know that your creative brain was on the job.

It's almost as if we have some sort of a "keep an eye on the store" ability that both monitors the satisfaction level of our basic needs and offers us help in satisfying them, whether we want it or not. When we were insulted, our need for power was frustrated and our brain went right to work to remedy the insult. In this case it was a little late but we still appreciate its effort. But when it offered you a clever retort, you didn't think there was anything wrong with your brain. If you shared that comeback with some friends, they'd have enjoyed it, no one would have called you crazy.

When it shows itself, especially when you are unhappy, your creativity is not usually as helpful as in the example above. More often it shows itself, either to you or to the people who are concerned about you, in ways that are mysterious, painful, and frightening, such as hallucinations or mania or even as painful, disabling autoimmune symptoms that may mysteriously flare or go into remission. More often than we would like, it offers us creative but self-defeating choices, such as to get behind the wheel and drive when we are drunk.

Suppose, instead of your creativity presenting an idea to you as a thought, it created a voice uttering a threat or any other message directly into the auditory cortex of your brain. You would hear an actual voice or voices; it could be a stranger or you might recognize whose voice it was. It would be impossible, just by hearing it, for you to tell it from an actual voice or voices.

You might look around to see who was talking to you but see no one. But the voice could be so real you might still think if it wasn't a person, it was coming from somewhere, maybe out of the lightbulb in the ceiling or from a neighbor's television antenna. If you told anyone you were hearing a voice talking to you, that person would likely conclude you were crazy and needed help. If you told this to an establishment psychiatrist, in a few minutes, you'd be diagnosed as having schizophrenia, told that there was pathology in your brain and that you needed strong brain drugs to stop the symptoms. If you seemed too out of contact to accept the psychiatric intervention, this would be told to your family and the treatment might be forced on you.

I accept it is abnormal for you to hear a voice that you created. But it's almost infinitely more abnormal, in terms of statistical probability, to think creatively enough to do the work that led to a Nobel Prize, as John Nash did.[1] It is, however, less improbable for such a creative mind as his to go on and create the hallucinations and delusions that led him to be diagnosed with schizophrenia in 1959 and to continue to carry the diagnosis for over twenty years. But by 1993, he was sane enough to be awarded the Nobel Prize. As of 2002 he continues to be sane.[2] Attention-getting creativity, both sane and insane, does

---

[1]John Nash, born in 1928, did the creative work in the 1950s that led to his winning the Nobel Prize in 1994. What makes what he did so unusual is that he was diagnosed as having schizophrenia in 1959. He stands as an example that schizophrenia is not a mental illness, but a series of symptoms such as hallucinations and delusions. There is no damage to the brain and many people recover from it completely, as he evidently has. His story was glamorized in the 2001 movie *A Beautiful Mind*. For a more accurate version, see Sylvia Nasar, *A Beautiful Mind* (New York: Touchstone Books, 1998).

[2]What I have just explained was documented on April 28, 2002, on a program entitled *A Brilliant Madness*, broadcast on Public Television at 9:00 P.M. Nash, a highly creative mathematician, participated in the program and explained how he willed himself back to sanity. My belief is insanity was no longer working for him.

not indicate mental illness. It indicates an unhappy brain doing what it is capable of doing.

It is hard for anyone, even for a psychiatrist, to conceive of the creative potential of our brains. If you can hear voices, your brain can create voices. If you can see, it can create visual hallucinations. If you can feel pain, it can create pain, often in greater duration and severity than you would ever experience from an illness or injury. If you can feel fear, it can create disabling phobias. If you can think rationally, it can go beyond rationality and create supernormal mania. No one can predict what a brain can create. But it is safe to assume that no matter what total behaviors you may experience or see around you, your creativity or someone else's may be the source.

But as startling as some unhappy creations might be, the vast majority of what the brain creates is helpful to you and others. Think of a world without creativity: a world without music, art, literature, science, humor, innovation, or athleticism, and you'll begin to realize the creative potential of the four hundred billion neuronal connections in its gray matter.

Think of your creativity as a wild card in your brain. It is capable, on its own, of creating anything you are capable of experiencing and almost anything you or anyone else can conceive of experiencing. Creating a clever retort, a hallucination, a phobia, a severe backache, or even an autoimmune disease is child's play for your creative brain. It is our creativity that provides us with what I and many others believe is our humanity.

What is so unfortunate is that with all the wonders of our creativity, we have also created a dead, totally noncreative environment populated by people just like you and me but with one big difference: their brains have made the mistake of creating the symptoms of what is called mental illness.

These dead environments are growing and are occupied by thousands of people diagnosed with symptoms such as hallucinations, delusions, and mania and treated with what are called the major tranquilizers. Under the effect of these powerful drugs, they are, indeed, tranquilized. Almost every creative possibility in their brains can be eliminated by large doses of this medication. They may no longer hear voices but they have also been robbed of what makes them most human, their creativity.

Suppose you started to hallucinate, certainly a strong indication of how unhappy you are, and instead of brain drugs you were treated with kindness and support by your family and friends. And also by psychiatrists and other mental health professionals who, while aware of your unhappiness, were not frightened or tempted by it to intervene with drugs. Those professionals would use their expertise to reach you and lead you back to reality.

They would explain what is going on to your frightened family and help them to realize you are lonely and may feel no one respects you or understands you. They are prepared to provide you with gentle, noncontrolling companionship and respect and to counsel your family to do the same.

Mostly, they would not try to control you by telling you that you *have* to change the way you are thinking or acting, and they would counsel your family and friends not to do this either. They would tell the people who are close and want so much to help you not to use any external control in the way they deal with you. They would avoid telling you how talented or gifted you are and how you are not taking advantage of your gifts or talents. Or they would be careful to avoid sending you this same message indirectly through looks and implications.

These high expectations coupled with messages urging you to do a lot more with your life than you're doing are difficult for many young people to deal with and may have something to do with the outpouring of creative thinking that is now labeled schizophrenia. Others may cope less creatively with this pressure by turning to addicting drugs. In both instances, the person with the symptoms may be trying to escape from both the message and the messengers with creative symptoms or drugs.

Once a family member gets so creative that he seems to have lost touch with reality, the people who can help him, family and professionals, should try not to use any external control; they should not query him about his hallucinations and delusions, and should be ready to change the subject if he tries to talk about them. They should try to keep close to him and attempt to create conversations with him that are perfectly normal on their side no matter what he says.

Basically, no matter how crazy a person is, the people who care about him should try not to get involved in his craziness. When I was dealing with people suffering from schizophrenic symptoms in the Brentwood Veterans Hospital in West Los Angeles, I did my best to relate to them and, especially, to get them involved in any sane behavior I could share with them. I found that the best way to do this was to ask them to join with me in sweeping the floor. I'd approach a patient on the ward with a broom and a dustpan. In those days they all smoked and the floor was always filled with butts.

I'd introduce myself as the patient's doctor and say, "Would you rather sweep or hold the dustpan while I sweep?" I rarely got refused and we swept up the floor slowly and carefully. It was important for their self-respect that we did a good job. As we swept, I began to tell them how important it was for us to get to know each other. After we swept, I'd suggest shooting a game of pool or even watching television together. When they would go to one of the hospital activities, such as art, music, or any of the many other therapies the hospital offered, I'd join with them and continue to talk.

What I did was try to help them satisfy their need for love and belonging with me in a completely sane way. What I wanted them not to do, as I suspected many of them were doing in their craziness, was to reject me before I could reject them. My approach was *I accept you but I have no interest in your craziness. I will protect you from harming yourself and others, all the while encouraging you to do anything you want to do that you are capable of doing.* This is what I would advise Selma to do with her son, accept him as he is and get involved in any sane activity both of them can enjoy. No matter what he does or says, she should keep her side of it sane and noncontrolling.

As I dealt with psychotic people without drugs (giving them drugs would be no different but less effective) I talked with my mentor, Dr. G. L. Harrington, about how most of what our patients did was sane or close to it. They dressed, bathed, shaved, ate, went to the bathroom, watched and understood television, participated in the many hospital activities, and knew what they were doing. Their talk or attention wasn't always sane, but what they did was mostly sane and, through these activities, I, the aides, nurses, and therapists could reach them.

The main message was that you are now safe and cared for. Drugs and talking to them about their craziness sent exactly the wrong message. It was a cardinal rule never to mention the word schizophrenia to them. We were encouraged to keep in mind that all of them had been sane at one time, that being sane wasn't something new. It was the insanity that was new, and ignoring it as much as possible sent them the message that it was now safe to go back to the way they had been. Communicating this to them by talking and working with them realistically was something the family and friends who understood them and their creativity could learn to do.

As stated in the first chapter, there are thousands of people with serious *DSM-IV* symptoms, including the symptoms of schizophrenia, who by luck or good sense, are treated this way by professionals, family, and friends. When they are able to regain the sane connections they once had, the symptoms tend to disappear.

But there are also millions of people diagnosed with a new condition (1987) called fibromyalgia, who have physical symptoms such as aching, paining, insomnia, fatigue, muscular weakness, abdominal bloating, and urinary problems, but no physician, using all the diagnostic tools available, has been able to find any pathology to explain their suffering. Fibromyalgics feel a lot of pain but are still normal in the way they talk, think, and relate to the people in their lives. Added to the people with migraine headaches, low back pain, painful joints and chronic gastrointestinal upset, you have millions more people suffering chronic pain with no pathology to explain it.

Their problem is that most of the time neither they nor their doctors are willing to consider the possibility that these created symptoms are the result of unhappiness. What almost all the sufferers and many of their physicians are more than willing to accept is the idea that they are physically ill. Even though no pathology can be found, it is hard for doctors and almost impossible for patients to give up their belief that there is something physically wrong with their bodies that some day may be found.

For many people diagnosed with fibromyalgia or other chronic pain, the admission of unhappiness caused by unsatisfying relationships is more destructive to their self-esteem than the pain of their "disease." For many of their doctors, pain is one of the most acceptable of all symptoms and must be treated with drugs even though there is no pathology and it does not respond to any nonaddictive pain drug. The connection between unhappy relationships and pain is almost as difficult for their doctors to accept as it is for them to accept. The unwillingness of the medical profession to come to grips with the creativity of an unhappy brain costs billions of dollars every year. If we wait for the medical profession to take the lead here, we will wait forever. My goal in this book is to bring the idea of total behavior and creativity directly to the public. People have a lot to gain and nothing to lose by getting rid of external control and replacing it with choice theory. In the end even the doctors will gain a lot. They will be free of people they can't treat medically and have more time to treat those they can help. The drug companies have a lot to lose but I am not worried about them. There are plenty of people who need medication for actual pathology to keep them in business.

Finally, for Jeff with arthritis, the concept of total behavior may help him understand what now is a medical mystery: autoimmune disease. In rheumatoid arthritis, the immune system does what it should never do: it attacks and destroys the normal cartilage in his joints. While doctors know what happens, no one yet knows why this happens. There is no curative treatment for Jeff. All the treatments, drugs, and surgery are palliative.

But suppose Jeff was willing to accept that he was unhappy when his immune system started to become crazily creative. Unlike most harmful creativity, which does not do physical harm to the body, rheumatoid arthritis does. Since Jeff has complete control over his thinking and acting, he can begin to help himself by stopping his attempts to control any of the important people in his life. He can also try to protect himself by explaining what he is doing to these people and asking them to stop trying to control him. If, by doing this, he can improve his relationships and become happier, there is a chance his immune system will return to normal.

There is a precedent for this in Norman Cousins's best-selling book, *An Anatomy of an Illness,*[3] I also have had some positive experiences with this educational approach to rheumatoid arthritis, which I explained at the end of Chapter 3.

Restraining Angering with Depressing and Other Symptoms

Finally, I'd like to deal with the utility of the symptoms people create as in psychosis or learn as in depressing. They can be useful to the creator in at least three different ways besides warning him to do something about his unhappiness. Painful as they are, they can prevent him from becoming even more unhappy by helping him to restrain his anger. For many people, *angering* needs to be restrained. Unrestrained angering can lead us into more unhappiness than anything else we can possibly do.

For example, recently two young mothers, one in Illinois, the other in Texas, killed their young children: three in Illinois, five in Texas. Both women were declared sane, found guilty of first-degree murder, and have been given life sentences. Without going into detail, I can tell you that these were very unhappy, very angry women.

It is not my place to comment on the fairness of the trials or the verdicts. I do not have all the information. But I can declare, based on many years of experience, that when people are unable to restrain their anger they are capable of doing immense harm. I also believe that none of us can avoid angering when we are frustrated. This is so because angering is the only feeling we are born with. All other feelings are learned.

Newborn babies are angry as little hornets when they dis-

[3] New York: W.W. Norton, 1979.

cover how much less comfortable it is out here than it was in the womb. Unlike depressing, children don't have to learn to anger from the people around them; it's built into their genes as a basic survival behavior such as swallowing. It gets them the attention they need to stay alive. A couple of years later, when they arrive at what are called the "terrible twos," they deal with almost every frustration by angering, even tantrumming.

But sometime after two, almost all babies discover that tantrumming doesn't work. Their mothers just let them have at it. By then both their ability to learn from their surroundings and their ability to create kicks in and they start to learn how to depress. Depressing, however, is not a genetic behavior; I believe we all learn how to do it from people around us. There is no shortage of role models for this behavior. Without depressing our society would suffer from much more murder and mayhem than we have now.

Depressing is almost the exact opposite of angering. Angering is quick and unrestrained. Depressing is slow and very restrained. Angering knows little reason, depressing is filled with reasons. Because there are so many unhappy, frustrated people who anger a great deal, our society is filled with individuals who have chosen to depress to keep their anger restrained. But if the situation in which they are depressing continues, they may not be able to depress strongly enough to restrain their anger.

I am sure that before they killed their children, both women depressed, and after they killed them, they depressed even more. But in both instances, they were not able to continue to depress strongly enough to restrain the anger that overwhelmed them in these two situations. You can see similar stories in the papers or on the television every day of the year. Oklahoma City, Columbine High School in Littleton, Colorado, the World Trade Center, the high school in Germany, are social and political examples of what can happen when angering is unrestrained.

Unfortunately, psychiatry has many drugs for depressing and few for angering. This is because most people begin to depress as soon as they feel anger. The number of people who depress may outnumber the people who anger enough to harm another person, ten thousand to one. If the number of people who are now depressing, and in doing so restraining their anger, stopped depressing and returned to angering, our society would fall apart.

In the instance of the two women, it is possible that they were on antidepressant drugs that made it hard for them to depress and restrain their angering. Doctors who prescribe these drugs are told to warn patients and their families about the possibility that they may burst into anger on these drugs or if they have been suddenly taken off them. Going on and off drugs that affect how we feel—alcohol is such a drug—may make it more difficult for us to deal with frustration by restraining anger.

When we are small, angering becomes less and less effective. The adults are too powerful and we look for more effective ways to restrain the anger than to depress. It's not that we stop depressing; it is still the total behavior most widely chosen to restrain angering. But as we create other symptoms when we are unhappy, we discover that almost all of them will restrain angering.

Two of the most kind, supportive women I have ever attempted to counsel were both suffering from severe rheumatoid arthritis. I felt then, and I still feel this way, that these unbelievably genteel women, so accepting of their disease, were using their arthritis to restrain huge amounts of anger in one or more of their frustrating relationships. For some reason they did not want to depress, or maybe they thought that they would not be able to depress strongly enough to restrain their anger. It seemed to me that I could feel their anger right through their arthritis.

In both instances, when I tried to get them to talk about their relationships, they refused. I had only one consultation with either of them. They thanked me but said that they were not interested in more counseling. This happened before I had developed the choice theory I've been explaining. If I had a chance to talk with them now, I might have the skill to persuade them to enter counseling. I feel the same about women with fibromyalgia. They create the pain to restrain their angering.

Besides Restraining Anger, There Are Two Other Ways We Use Symptoms, Such As Depressing, When We Are Unhappy

Early in life we also learn that all the symptoms, painful, uncomfortable, or crazy as they are, can serve another very useful purpose: *They are a powerful way to ask for help, even to control another person, without begging.* Asking and, especially, begging for help frustrates our need for power. No one respects a beggar. Of all the symptoms, depressing does this best. When we choose to depress, those around us reach out to us. But other symptoms, especially suffering pain or acting crazily, may also serve us well.

Specifically, depressing, if we are willing to put up with the misery, is a very effective way to control the people around us, but usually for only a while. If it goes into many months or years, we have to increase the depressing to get the attention we want. If we don't repair the relationship that is not working for us, we may depress for years. All the other symptoms work in the same way. We tend to choose the one with which we are most familiar or we see others use effectively. Or we tend to create new ones such as craziness that get us a lot of attention and control. But exactly why we choose one over another is not yet known.

The other reason we choose to depress or use other symptoms is to help us *avoid doing something that we believe will cause us more pain than we are already feeling.* We then can say we're too upset to do it or we're unable to do it because the symptom is so disabling we can't do it. Examples of such symptoms are panicking, obsessing, compulsing, or phobicking. I'll give you a common example of what I mean. It can happen to men as much as to women.

Suppose you are a woman who has been involved in a long relationship that you believed would end in marriage. But after a number of years it falls apart; the marriage offer you counted on has been withdrawn. I'm your friend and I try to comfort you by suggesting you ought to get out and get involved in some social activities where you might meet a new companion. But as soon as I suggest this, you turn me down by saying, "I'm far too depressed to try anything like that. If I meet someone the way I feel now, I'll mess it all up."

A few months go by, and I try again and still you turn me down. It doesn't seem to me you're depressing that strongly but you're adamant. Then, because we're very good friends, I confront you on your continuing refusal to try to get socially reinvolved. You admit to me you want to, but whenever you get the idea you ought to do something, a wave of depression (your term) hits you. I do my best to explain to you that you're so afraid of another rejection, you're choosing to depress to avoid that possibility. I say to you, "Please I'm only explaining what's happening. I'm not trying to push you into anything you don't want to do."

You start to cry and admit I'm right. You ask me how you

can break out of this bind. I advise you what I always advise people involved in this painful choice. The first thing I say is, don't be too proud to admit to yourself that you're scared. Make admitting your fear a part of the way you reinvolve yourself in the dating scene. Do it first in groups where you can avoid an intense relationship. But then if you meet someone, experience the fear, and start to depress, admit immediately what's going on. There's a good chance that the man you meet is afraid of the same thing.

Fear of rejection is one of the most common human fears. Depressing keeps that fear in check. But if I were a woman or man reading this book, I'd make every effort to keep any new relationship as free as you can from external control. Most likely she and the man she is involved with could try to do what Joan and Barry did: rid their relationship of external control. If you are reentering the social scene, don't repeat what may have destroyed the old relationship. Joan and Barry used all three of these reasons as they put themselves first and their marriage second.

I'll send this chapter along to Joan and Barry and wait to see who comes for the fourth session, which will, finally, be the first large Choice Theory Focus Group. If they all come it will be Joan; Barry; Bev, who depresses; Jill, who migraines; Jeff, who has rheumatoid arthritis; Selma, whose son has been diagnosed with schizophrenia; and Neil, the man, whose sister suffers from panic attacks.

ten

The Fourth Choice Theory
Focus Group Session

Joan called to tell me she had passed the first nine chapters around to everyone in the book club who was interested in reading the material. I was excited about the possibilities of meeting with this group; they were used to talking about what they'd read.

Three weeks later, she called me again and said, "Last night we had our monthly meeting and it never seemed to end. After the regular meeting was over, all the people who'd read the nine chapters wanted to stay and talk about them. They asked me if they could all come to the Choice Theory Focus Group, so I have to ask you if it's okay. And by the way, Neil, he's the man whose sister has panic attacks, had her there. She'd read the material I gave him and she wants to come, too. What do you think?"

When I counted up the people, there were nine plus Carleen and myself, too many for my small office; we'd use our home. Since I have to record the meeting to write it up, we'll use our large oval dining room table. The recorder is very sensitive; if I put it in the middle it'll pick up everyone. I told Joan,

"It's fine. We can meet at our home. I'll leave it up to you to work out the details. Okay?"

Joan was excited, too, and said she'd handle everything. A few days later she called again and told me that another woman in the group has fibromyalgia and she's very anxious to come. "She's gone onto your Web site and found out you've written about it.[1] Can we squeeze her in?" I told her no problem, and she said, "Can you believe it? We could fill a whole psych ward from that book club, but when we meet, everyone seems okay."

"They are okay. They're a group of normal people with some unhappiness in their lives. When you're unhappy, you have symptoms. They're no big deal unless you go to a doctor who makes them into a big deal by calling them a mental illness. Actually, they're just the kind of people I was thinking about when I decided to write the book. I look forward to meeting with them. Joan, thanks for all you've done. And thank Barry, too."

After some negotiations, Joan decided an evening meeting would work best. I was disappointed that after all the work Joan had done she couldn't come on the only evening everyone else could. She'd planned to spend it with her mother and she didn't want to cancel. She said she'd attend the next session, she'd listen to the tape and Barry would fill her in.

Since this is study and discussion, not therapy, I won't attempt to describe any of the people who attended or give you any of their backgrounds except for telling you, as I already have in Chapter 1, what brings them to the focus group. Anyone getting a focus group started at an HMO or a mental

[1]See Chapter 14 for a list of all the books I've written that support the material in this book. The book I've written on fibromyalgia is called *Fibromyalgia: Help from a Completely New Perspective.*

health association wouldn't even have this much information. Teachers don't profile their students, they teach what they believe, encourage discussion, and answer questions.

When I teach, I like to be on a first-name basis, so I asked the group to call us Bill and Carleen. The group squeezed around our dining room table, the closeness providing an immediate sense of connection. Carleen and I introduced ourselves to the group and decided we wouldn't need name tags because they knew each other. Even Amy, who attended the book club meetings once in a while with her brother, was not a stranger to this group.

In the material that follows I'll mention who's talking unless it's obvious. If there is no mention of who's talking, it's I.

I started the meeting by saying, "Since this is strictly study, not therapy, I don't need to know any more about you than any teacher would: your name and what you look like. But since you know each other, feel free to tell the rest of us anything you want to about yourself, why you're here, or to ask questions and make comments about what someone else says. Carleen and I are here to encourage discussion and answer your questions. Okay, I'm looking for a question to get us started. Bev, you seem ready to say something."

"I don't know anything about mental health, but when I'm depressed, I feel like something is wrong with my brain. How can you be so sure there isn't?"

"Personally, I'm pretty sure. I haven't believed in mental illness since I was a psychiatric resident back in the fifties. But that's not the point. I'm not asking any of you to give up what you believe or talk you into anything else. You're here because you've read my material and you're interested. I don't mind if you have some doubts about what I've written but I'd appreciate it if you'd tell me why you decided to be here? Bev? Anyone?"

Barry answered, "I'm happy to tell you why I'm here. I'm here to learn how to be happier. Also, Joan and I are skeptical about your choosing-pain idea. I've read what you wrote in Chapter 9 and I have some questions I'd like to ask. I'm sure that Jill, Molly, and Jeff might have a few questions about that, too. But mainly I'm here because what you taught has helped our marriage and it has helped me at work. I want to learn more. Bev, why are you here?"

"Because I'm miserable. And if you're right and I'm choosing to be miserable, I'm even more miserable. I'm here to get help, not more upset."

Jeff said, "Look at this arthritic hand. Look at it! I can't even open it up and it hurts all the time. I've been seeing doctors and taking medicine for almost fifteen years and it's a lot more than my hand. I was seventeen when this thing hit me and it's getting worse. Fifteen years and all they've told me is I have rheumatoid arthritis and I may never get rid of it. Before I read what Bill wrote, no one had ever suggested I could do anything to help myself. He hasn't asked me to stop my medication or to stop seeing my doctor. All he's asked is we learn a little choice theory. What have I got to lose? And I've been thinking a lot about unhappiness. I thought I was happy before I got sick but now I'm not so sure. Tell me, Bev, before you got depressed, were you happy?"

Selma piped up a little sarcastically, "Jeff, pay attention. The question you should have asked Bev is, were you happy before you chose to depress? My son was a happy kid just starting college, when he chose to become schizophrenic. You did say he chose that, didn't you, Dr. Glasser?"

"Please call me Bill. But no, actually I didn't say he chose his psychosis. Does anyone know what I did say? I mean in what you read."

Jill answered, "I know exactly what you said. I read those

chapters very carefully. You said that our symptom, or symptoms, are part of a total behavior we choose when we're unhappy. That means all of us were unhappy before our symptoms came. Some of our symptoms, like depressing, we learned. The rest of them, like craziness, pain, or inflammation we created when we were unhappy. That also means that whatever we were choosing to do before we chose the total behavior, which includes our symptoms, didn't satisfy our basic needs. That's what I think you said in those chapters. Bev, if you don't accept the idea of total behavior, you're going to have a lot of problems with Bill's ideas."

Bev said, "Jill, do you accept Bill's ideas?"

"I'll be happy to tell you. A week ago, in the evening, just after I finished reading Chapter 9, I had a migraine headache. It didn't feel like I was choosing it. But as soon as I felt it coming on I said to myself, am I unhappy right now? And I was, I was very unhappy all that day. It's not that I'm happy that much but this was a really bad day. I'm a doctor, a family practitioner. I work for an HMO and I'm told what I can do and what I can't do. Some bean counter with eyes on the bottom line turned down my request for a patient to see a specialist. I don't ask for that very often, so this was unexpected. I was mad as hell but then I started having a headache. It's like Bill said, if I hadn't come down with that headache I might have gone to that guy's office the next morning and strangled him. But since I read that material, I look at the bright side; I've only got headaches. Ten women in my practice have fibromyalgia; they hurt all over. Ask Molly about that. [Molly nodded when she said that.] If I can get some of this anger under better control, I may get rid of my headaches. We're all unhappy; we just deal with it with different symptoms. Bev, what are you so angry about?"

Barry joined in, "I used to anger all the time and Joan

depressed to deal with her anger. Jill, say angering or choosing to anger instead of just the one word *anger*. Talking the talk seems to help."

I said, "I think Jill said something important. What really bothered her was that she thought she had a chance to get what she wanted when she called that case manager. Maybe there are two kinds of unhappiness. There's unhappiness when you think you should be able to do something about a problem. Like you get your hopes up and then you feel powerless when you get turned down. And there's unhappiness that you can't do anything about, like the tightwad policies of her HMO and your need for power is also frustrated. Either way you anger a lot."

Amy said, "I see what you mean. I've been thinking about my panic attacks a lot since Neil told me about your ideas, and I've read the material carefully. They seem to come right out of the blue like Jill's headaches. Maybe they're not out of the blue but I'm not ready to go into my personal business. But I'm more unhappy when I think I should be able to do something about it and then I find out I can't. It's like I think I have control over the situation and then I find out I don't. Maybe that's when I panic, to cover up the anger."

Molly said, "Bill said we choose the symptoms to restrain our anger. But when I start to hurt, pain takes over my whole life. I don't think about angering. Okay, I can call it angering. But all I think about is how can I get rid of the pain? It's awful. I can't even walk. It blots everything else out."

Amy said, "It's the same with me. I don't feel angry. Before I read those chapters, all I could think about were my panic attacks, like all Molly thinks about is her pain. Also, I felt like what I'm sure Jeff feels about his arthritis. I'm unlucky. I'm here because no one has been able to do anything for me. Now, after reading what Bill wrote, I'm getting the feeling I could do something for myself. Neil's been telling me I have some control over these attacks. He pointed out I never have them while I'm driving, and I drive a lot. But when I drive I feel in control, at least in control of the car. I've got to do a lot more thinking about external control and angering."

Selma said, "I've got a son with schizophrenia. He can't take care of himself."

Jeff interrupted, "You mean he can't get dressed or feed himself; he doesn't know what's going on, like on television?"

Selma said with some irritation, "He can take care of those things; he's not completely out of it but I don't think he could take care of himself if he didn't live at home. He moves so slowly on those drugs they give him that I lose patience and do things for him. He doesn't seem to hear the voices so much now that he's been on drugs but he doesn't seem to hear much of anything. He's still out of it. The drugs haven't helped nearly as much as I hoped they would. I've been taking care of him for ten years. I've tried and tried. I've done everything. I've read all the books. I'm a member of a support group for parents of children with schizophrenia. We all try; that's all we do, we live for our children. I had a son and now he's gone. I want him back to the way he was before he got sick."

"Selma, you said you've tried and tried. Did anything in what you've read suggest that you could change your son or anyone else?"

Selma answered again with some irritation, "That's all you talked about, people helping each other and improving their relationships. I read it; Barry helped Joan and Joan helped him. I do all I can to have a good relationship with my son. I get the feeling that lots of times he doesn't even know I'm there. But I know I'm there. I'll never give up."

I said, "Barry, help Selma out. Did you help Joan? Did Joan help you? You both got some help. What happened?"

"No, it's like you said in the book. In my mind I thought I was helping her by trying to get her to change. I used all the deadly habits; I criticized her and blamed her. And all she did was depress and stop talking to me. Then we decided instead of trying to change each other, we'd try to change ourselves. We put all our effort into our marriage and things got better. Selma, I'm sorry Joan isn't here; she had a date with her mother that she didn't want to cancel. Maybe she could explain it better. We've even stopped trying to change our son, Josh. We've been getting along a lot better with him at home and he's no worse at school. If I were you I'd stop trying to change your son. I may be speaking out of turn, but I don't think he wants the kind of relationship with you that you want. You've been trying for ten years and you tell us that a lot of the time he doesn't even know you're there. I'd back off for a while if I were you. It's all there in Chapter Nine; read it again."

Selma said, "But you and Joan are normal. Jim is sick. You don't understand."

Barry answered, "For the last ten years of our marriage, there was nothing normal about Joan and me unless you call being miserable normal."

Selma said, "Barry, Jim is mentally ill."

Jeff said, "According to what I read in the book, he's not mentally ill. Jim's like me, unhappy but creative."

Selma raised her voice at Jeff and said, "You're normal and he's sick. He's nothing like you?"

Jeff said, "Look at this hand again, Selma, does this look like a normal hand. If I could trade what I've got for a little schizophrenia, I'd sure give it a try. I hurt all day and all night like Molly and I'm crippled besides. Is Jim crazy all day long? Like he never has a normal moment? I doubt that very much. Your son Jim and I are both unhappy. I don't know about Barry and Joan but everyone in this room is unhappy and I'll bet we're all caught up in external control. We use it and we try to escape from it being used on us. I think his schizophrenia is his way of trying to escape from something. By now it may be your control. Love him even when he's a little crazy is my advice."

Selma got more angry and said, "So whose control are you trying to escape from with your arthritis?"

"I've been thinking about that. I've been thinking about it for three weeks since I read the last chapter."

After Jeff's comment and Selma's challenge the room got quiet. Then Carleen said, "Molly, you seem to want to say something."

"I don't know what to say. When I read the book I saw myself over and over. I said to myself, 'Molly, you're a control freak. Your whole life is based on control.' I try to control my husband and my children but even more I try to control myself. I have to be the perfect wife, the perfect mother, the perfect homemaker, the perfect gardener. I've even learned to groom our two little perfect poodles. I'm never satisfied with myself. I've been thinking and thinking since I read those chapters and you know what I figured out? [We all looked at her.] Now that I have fibromyalgia, I'm trying to be the perfect patient. But it's hard because my doctor doesn't really tell me what to do. Okay, Bill, I'm willing to agree with everything you wrote. But where's the chapter that tells me what to do? Tell me and you can count on me to do it perfectly."

I said, "Molly, you know what to do. Everyone in this room knows what to do. It's in the book. Relax your external control. Ease off on others and please don't forget to ease off on yourself, too. I guess I didn't make that clear enough in the book."

Jill added, "In my practice the ten women like Molly are all trying to do it all perfectly. But I also see a lot of people with arthritis like Jeff. They put all their effort into their disease;

they are really good patients. That struck me when Molly said she was even trying to be a perfect patient. It kind of fits me, too. I have high standards but working for the HMO I can't even be a good doctor much less a perfect doctor. But the headaches started a long time before med school. They started in college and I was happy then. At least I thought I was."

I said, "Most premed students worry about getting into medical school."

Jill said, "You're right about that. I'm surprised it took so long for the headaches to start. I've been a perfect student since kindergarten. Now I put too much time and effort into hating the HMO. I'm going to start putting my creativity into my job instead of into my headaches."

I said, "It was tough in school but if you studied hard you had some control. Now when you work hard and ask for a specialist, your needs are frustrated and you lose control."

Carleen said, "Bev, you've heard what everyone's been saying. How would you answer Jeff's question about how so many of us are involved in too much control? I mean both controlling ourselves and others."

"What do you mean?"

Jeff stepped in, "Bev, she means what are you unhappy about and how much external control is involved?"

Bev answered, "I don't know about the external control part unless you call trying to be a good mother external control."

Selma said, "Bev, that's just what they've been saying. Like because I take care of Jim, I've driven him crazy by trying to control him too much."

Jeff jumped in to say, "I'm sorry, Selma. You didn't have anything to do with your son being what he is any more than my mother had anything to do with my arthritis. Believe me, she's as concerned about me as you are about Jim. But I've been thinking. Maybe all that care and concern was more than I could handle. All I'm saying is ease off a little. Then he won't have to worry so much about you being disappointed in him."

Barry said, "Okay, Bev, you're very concerned about your kid. I can relate to that. Do you want to tell us about it? Maybe it would help to talk about it."

"I have a seventeen-year-old daughter who drives me up the wall. You claim I know what to do to get along better with her and you're right. All I have to do is let her stop doing schoolwork and party all weekend. Because I'm trying to keep her from ruining her life, this girl says she hates me. You should hear how she talks to me. I'm beginning to believe she really does hate me. I'm depressed and Bill says I'm choosing it. What good does it do to tell me that? She doesn't care how I feel. I think she's happy when she sees how upset I am."

I became a little caught up in trying to defend myself and I said, "All I've been trying to teach you is that the problem between parents and children is external control. Good parents are as much into it as neglectful parents. You're using it on her and she's using it on you. You both hate it but you both keep using it. You're depressing like Joan, and your daughter's angering like Barry. The more you depress, the more she angers, because she believes you'll finally get so upset you'll give up and let her have her own way. That's what she wants, isn't it? Her own way?"

Amy said, "That's what we all want, our own way. That's what everyone wants, isn't it?"

Roger said, "There's nothing wrong with wanting our own way. It's what we do when we don't get it, try to force others to let us have it, that causes so many symptoms."

Carleen said, "Bev, you're in a tougher spot than Joan and Barry. If they're unhappy, they can separate or get a divorce. You don't have that option with your daughter any more than

Selma has with Jim. Brandi may tell you she's out of there as soon as she turns eighteen but it's mostly talk. Even if she leaves, believe me she'll be back. You're going to be with her, fighting or loving, for the rest of your life. I say keep the loving and stop the fighting."

"Well, tell me how. I don't know what to do."

I said, "Bev, this isn't therapy. There are good counselors you could talk with who could help you with her. Roger's here; he could refer you to one if he's busy. But you could also consider doing what I suggested in the chapters we sent you. Do what Joan and Barry did."

Barry said, "Bev, knock off the deadly habits, they aren't working. If I could do it, anyone could do it. We've stopped talking to Josh about school, you know, threatening him and punishing him. If he doesn't do his homework, we keep our mouths shut. But we both spend a lot of time loving him and talking to him. It all changes after you stop using external control."

Bev said, "But that's crazy. Josh is a kid. Brandi is a senior in high school. If I don't keep after her to study and do her homework, she won't graduate."

Jill said, "I've never had any kids but my guess is if you keep after her, you'll make things worse."

Roger said, "We had some problems like that with Joan when she was fifteen. It was in Chapter Seven. We loved her and didn't try to control her. We even loved Barry who was having sex with her every day. It wasn't easy but we did it."

Bev said, "If I didn't love Brandi, I wouldn't be upset."

Barry said, "It's too bad Joan isn't here. It'd do you good to talk to her. She gave Roger and Jean a real hard time and I sure didn't help. But she always felt welcome at home and, it was weird, I did, too."

Molly said, "So why am I so screwed up? I was a perfect daughter; I still am. I did 'A' work right through college."

I said, "Molly, why are we the way we are? That's a real good question. I'm going to try to answer it in the next chapter. It's the only part of choice theory I haven't explained."

Barry said, "But Bev has a point. Josh is twelve. If he doesn't do his homework when he's a senior, I'm going to be upset. Homework's the killer in school. Except for Molly and Jill, did any of you do your homework?"

Selma said, "Jim did his, he was good in school until he got sick. I'd go over it with him all the time. The thing is he's still doing it. He scribbles a lot of crazy things that don't make sense. He shows it to me and tells me it's his homework."

Jeff said, "Selma, what do you do when he does that?"

"I used to pay attention to it. But when I read what Bill said in the last chapter I said to myself, I'm getting involved with his craziness and I shouldn't do that. So now I've been trying what Bill did in the Veterans Hospital. When Jim shows me his 'homework,' I ask him if he'd like to help me to do a simple sensible chore like go to the store with me. Not a big supermarket but a little neighborhood grocery where they know him. They talk to him, and most of the time he's sane with them. It seems to be working. He seems more able to shift gears from crazy stuff to the sensible thing we're doing. And I'm doing something else. Bill said never to mention the word *schizophrenia*. I never made that mistake; I hated that word. But what I'm trying to do now is to stop thinking that word. Whenever it comes to mind I say to myself, 'He's unhappy,' and I back off whatever I'm trying to do with him. Jeff, you ought to try it. When you think about your arthritis, think unhappy instead and ease off on yourself."

Jeff said, "I appreciate your suggestion. You sound tough at times, Selma, but you're really a very caring person."

When Jeff said that to Selma, I had the feeling the group was really connecting. Bev said to Barry she wasn't going to wait for the next session. "If it's okay with you, I'd like to call Joan and talk to her about Brandi." I can't exactly describe the group's mood but it was what I wanted, intense but still relaxed.

I asked myself, *What's different about this group from a therapy group?* One thing was obvious. Instead of fumbling around like most therapy groups do at the start, they had been given a focus by choice theory. It will have more of a focus when these study sessions are incorporated into the finished book.

During refreshments, the group stopped talking about their problems and about what they'd read. Instead they socialized. They made a real effort to get to know each other and they asked me a lot of questions about how I'd come up with the ideas of choice theory and who were my teachers. I talked a little about my father and how lucky I was to have a father that never used the deadly habits on me or anyone else. I told them about Dr. Harrington, my mentor, who connected with me when I told him I had real doubts about mental illness. That got them into talking about parents and they asked Roger how he and Jean came to be the way they are. With Roger it was his mother, who never used any external control. Then the group asked if his wife, Jean, could come to a meeting; they'd really like to meet her. He said he'd ask her.

This went on for over an hour. I wondered why they'd stopped talking about their problems. A few times I had an urge to ask them to get back to work but, fortunately, I kept quiet. It took me a while to realize that what they'd come for was more than to solve their problems. They'd come to relax and get to know each other. Now that they'd found out how important good relationships are, they wanted to get close to each other. Without any external control in the group, the air was filled with good feelings.

All my life I've noticed that when people get together to

learn and discuss ideas that are important to them, they also make an effort to get to know each other better. It's as if the more we learn, the more we want to get closer to the people we're learning it with. It's a supportive cycle. That's what was happening here in this break. But they were also experiencing something else that they enjoyed. I wasn't sure what it was but I was anxious to talk with Carleen about it after they left.

The break lasted quite a while, and we didn't push them to get back to business, and they never did. A few had to leave and then a few more. Finally, only Jeff was left. He wanted to talk with us, personally. He asked, "Do you have any hard evidence that my arthritis has a psychological component? I mean, besides your experience with that group. I've read the Norman Cousins book you referred to. I guess we all have. But this is a disease of remissions. I've had them for a while but then it always comes back. But do you know if anyone who's had the disease as long as I've had it has been cured?"

"Jeff, I can't answer that question because I can't answer it for unhappiness, either. I don't think there's a cure for unhappiness. But if you get rid of the external control you use and deal better with others who use it on you, you have a real chance to be happier than you've been. I think we're on the right track. This is the first meeting. Carleen and I worked with that group of people with arthritis nine months before most of them went into remission. All I can tell you is I think you made some very significant contributions. I hope you keep coming."

"Okay, I'm satisfied. Believe me, I'm not leaving the group. But it's okay that I asked you, isn't it?"

"Jeff, it's more than okay, I think it's progress. It's a pleasure to have you in the group. I appreciated your contributions."

Right before the first people left, I told them, "I'll get this chapter and the next one out to you in a couple of weeks. Read them over and we'll see you next month."

After Jeff left, Carleen was as anxious to talk as I was. She said, "They got along so well together. I realize they know each other from the reading group but I expected a hard time from Bev and Selma and some big arguments in the group. Why do you think it didn't happen?"

"I'm not sure they realized it but, except for a few words at the start, they didn't use external control at all in the session."

Carleen and I realized that that was what we have also noticed when we teach choice theory to teachers and counselors in our four-day training sessions. They keep telling us how much they enjoyed the training, how close they've become, and how much it's helped them. When I ask them if it's because we don't use any external control in our teaching, they agree. It's a new experience for them.

It began to happen about ten years ago, when we'd stopped evaluating the people who took our training. They'd hated it and it was pure external control. We thought we needed to do that to keep up the quality but, without it, it's been exactly the opposite. Our quality has gone up.

"We have to be real careful not to use it in our sessions," I said. "Bev was having trouble giving up external control with her daughter. We encouraged her to give it up but no one put her down because she was resisting. Bev was tense when we started; I guess she expected criticism. But she relaxed as the session went on."

Carleen said, "I was tempted to say something to her. We believe choice theory is so right for everyone, it's hard to teach it and still not use external control."

"I wanted to say something to Selma a few times, too, but I didn't. I don't think she would ever have said she tried my ideas if I'd said anything that even suggested control. It's funny how controlling people are so sensitive to control."

Carleen said, "This is why it wouldn't work if professionals took over these groups. Maybe that's why AA works so well. They do it for themselves and no one makes money off it. I'm sure some people in the group would want to be told what to do. That's what Jeff was driving at when he stayed to talk. Molly brought it up, too. She said, 'Write a chapter to tell me what to do, and you referred her back to the book.'"

"That's the difference between a book and a person. Even the kindest, most caring 'I know what's right for you' is still external control. The book's just information, it's totally devoid of control. What I liked is they kept referring to what they'd read. That's what I hoped would happen."

eleven

Luck, Intimacy, and Our Quality World

As a child, I was very lucky. I had an outstanding elementary school education. I remember all of my teachers' names and most of them I loved. But I loved my sixth grade English teacher, Miss Shehan, the most. Luck plays a big part in our lives. For me, being taught by Miss Shehan in the sixth grade was like winning the lottery. That year with her I did better in school than I'd ever done before. But it was not luck that led me to put her picture into my *quality world;* I recognized she was a very special person.

Our quality world sets the standard for our life. It is a small, simulated world that we start to create in our memory shortly after birth. By the time we are two years old, it is filled with need-satisfying information, mostly in the form of pictures, that we keep adjusting and updating all our lives. I can't bring Miss Shehan back to life, yet her picture is still in my quality world. When I sat down to write this chapter, she immediately came to mind.

Our quality worlds are composed of three kinds of need-satisfying pictures. First and foremost are *people* such as Miss

Shehan. Second are *things;* my computer is very much in mine. Third are *systems of belief* such as our religion and our politics. For me, choice theory is the major belief in my quality world. While people, things, and systems of belief are all important, it is the people we put into it who are most necessary for our mental health. They are in our quality world because we have a good relationship with them or we believe they are the kind of people we want, or would want, if it were possible, to have a good relationship with. Elvis Presley was and still is an important person in many people's quality world.

While we can become highly involved with *things* and *beliefs,* it is the ability to connect these things and beliefs pleasurably to the people in our lives that leads us to put them into our quality worlds. Becoming intimate with people is the way we satisfy all our basic needs, except for bare survival. Therefore, our quality worlds are filled with pictures of intimate experiences with people we care about and want to care about us.

Loving intimacy is the most obvious. This is sexual with a mate or lover, nonsexual with a child, parent, or close relative. But there is a less obvious intimacy, such as I had with Miss Shehan, which transcends sexual and family connections, that can be a very strong and positive aspect of our lives. It is between two people—two friends, a teacher and student, or two colleagues—in which each recognizes the other's worth but makes no effort to change the other. Filled with care and encouragement, it is completely devoid of external control.

Sexual intimacy and family intimacy are certainly pleasurable and begin easily. But over time, this intimacy becomes more and more difficult to maintain because the external control world we have grown up in begins to creep in. For too many of us, these relationships degenerate into *wanting* something from another or trying to *control* another to get it. This leads to feeling obligated, a feeling that always destroys intimacy. *Obligation* is external control.

With the exception of Joan, Barry, and Roger, I believe that all the people in the study group may be lacking the intimacy they want and need with at least one of the following: lovers, family, friends, and colleagues. It's not because they've never had intimate relationships; I'm sure they've had them in their lives. Rather, it's because they don't understand that intimacy cannot coexist with external control; they have lost these relationships and this loss results in the symptoms expressed in the group.

The intimacy I had with Miss Shehan can occur at any time, it can occur between people who have not been previously acquainted and it need not last very long. But when it happens, it's magical. Our confidence gets a boost; we feel empowered as if the world has begun to open up for us. The small incident that led to my intimacy with Miss Shehan may not seem like much, but spending that year in her English class had an effect on me that's lasted all my life. I've told this story many times, yet what happened remains with me as if it were yesterday.

It was early in the fall semester. I raised my hand, she called on me, and I answered the question I was sure I'd just heard her ask. She looked at me. I can't exactly remember the expression on her face, but as best I recollect, it seemed like a combination of surprise and curiosity but it was also caring. Her whole demeanor told me this was not the ordinary question-and-answer exchange between a student and a teacher.

Then in the soft voice that she used all that year—in fact I can't remember her ever raising her voice—she said, "Your answer is correct, Billy, but I haven't asked the question yet."

I remember being a little frightened as if I'd done something wrong. Immediately she saw the fear and confusion on my face and said, "It's all right, don't worry about it," and quickly went on with what she'd been talking about. I sensed she didn't want to get the class involved in this intimate moment between us. Right then I put her picture into my quality world and I believe she put mine into hers.

Nothing more occurred during that period and she continued to teach until the bell rang. As we were getting up to move to the next class, she asked me to wait; she wanted to talk with me. All the students quickly left and I went up to her desk where she was seated. Although she looked older than she was because of her prematurely white hair, she was young; my mother guessed her to be in her thirties. She was a beautiful woman with sparkling blue eyes. To me, she looked like an angel.

I could sense her acceptance but I was still a little apprehensive. This was a different moment with a teacher than I'd ever experienced before. I waited in front of her desk. I didn't say anything.

She asked, "Did you really hear me ask that question?"

"I thought I did. I'm sorry." I didn't know what else to say. I really thought I'd heard the question. As you know from Chapter 9, my creativity had inserted her voice asking that question into the auditory cortex of my brain.

She sensed my discomfort and quickly said, "It's all right. I think it's marvelous. If it ever happens again, please raise your hand like you did this time."

This interchange took no more than a minute and never happened again but I had a wonderful year in that class. I listened to everything she said as if she were talking to me, personally. I felt she also listened with great interest to whatever I said. With her interest and support that year, I did some very good writing, better than I'd ever done before.

I wrote an essay saying we should not prejudge teachers.

We should get to know them before we make up our minds. It was published in the school newspaper, *The Tattler,* and it led to some discussion at school and at home. My mother kept it for years but it got lost when she died. It could have been a prologue to the books I've written on education.

Another moment came when I had changed my career from chemical engineering to psychology. Since I'd had no psychology in engineering school, I had to take five undergraduate courses before I could move on to graduate work. In one course, the professor and dean of the college, Dean Huntley, asked me to stay after class. It was a small class, no more than fifteen students. We had time for a lot of class discussion and Dean Huntley got to know us personally. He asked me what my goal was, and I told him a Ph.D. in clinical psychology. He paused for a moment and then asked, "Why don't you go to medical school and become a psychiatrist?"

I told him I'd like to become a psychiatrist, I'd certainly thought about it, but my grades in engineering were too low. Besides I'd had no premed classes and I didn't think I had a chance of getting into medical school. He said I was making a mistake, I should try. I asked why he thought so and he said, "In the time I've gotten to know you, I don't think you will ever settle for being second best by definition."

As you may know, psychologists are lower in the mental health pecking order than psychiatrists, much lower fifty years ago than now. That intimate moment with Dean Huntley and our relationship in that class had a lot to do with my applying to medical school, finally getting accepted, and becoming a psychiatrist.

The final time it happened was in the third year of my psychiatric residency. I was explaining to my new supervising psychiatrist, G. L. Harrington, how I was dealing with a patient I was seeing. I took a chance and told him that I didn't believe there was anything wrong with her that warranted seeing a psychiatrist. She had been attending the clinic for three years as a teaching client telling the same story each year about how she'd been traumatized by her childhood relationship with her now deceased grandfather. None of the therapy had helped her but she enjoyed the attention she got by telling and retelling her story each year. She had been labeled with a series of *DSM-IV* diagnoses, usually a new one each year by the resident who saw her.

I explained to Harrington that she wasn't mentally ill; she was lonely and unhappy. She needed to be taught some different ways to get along with her husband and children. When I said this, Harrington reached over, shook my hand, and said, "Join the club." He didn't believe in mental illness either. It was another intimate moment. We stuck together, mentor and pupil, for the next seven years and with his support I got started in my unorthodox psychiatric career.

What happened with Miss Shehan has happened often enough in my life that I've become aware of my creativity. I'm not clairvoyant but in certain situations where people might ordinarily say or even think something pertinent to what we're talking about, I sense what's going through their minds. It's been a great help to me as a psychiatrist because sometimes I tell the client what I think she would like to say but may be too frightened to say it. I'm sure I'm not the only one who does this, but what surprises me is sometimes I sense their exact words, even seem to hear their voice, as I did in that sixth grade class.

That, however, doesn't happen very often. What does happen is that I try to send the clients I work with the message that it's safe to share anything on their minds: I won't judge or criticize. Anything you have to say could be important. I now realize that the way I send this message is to try very hard not to use any external control when I work with a client. If I slip, I apologize and use the apology to explain the external control I'm trying to avoid. As soon as I explain some of what I now call the deadly habits, they quickly catch on.

Intimacy, as it's understood by most people, is sexual. It's called intimacy when people who barely know each other, much less love each other, get quickly involved in enjoyable sex, often better sex than they have later when they are more familiar with each other. Familiarity may not breed contempt but it does breed external control. It's the seven deadly habits that take the intimacy out of most sexual relationships.

For me, intimacy, the experience of closeness, support and encouragement without control, is extremely pleasurable and need have nothing to do with sex, age, gender, religion, race, or station of life. Within the bounds of courtesy, it opens two people up to each other without either worrying that they have to watch what they say or do. It levels the playing field; there is no power differential between intimate people. It makes us immediately available to each other as it did for me with Miss Shehan, Dean Huntley, and Dr. Harrington.

It frequently occurs in the groups we teach choice theory and reality therapy to as the major part of the William Glasser Institute program. Carleen and I sensed it was beginning to happen in the study group depicted in the last chapter. The ability of an actor or actress to project intimacy is what makes them stars. Julia Roberts lights up the screen as soon as she appears.

In the last chapter, Molly asked why she was so involved with perfection and I told her to wait until she gets this chapter, which has to do with our quality world. I wanted the group to read this before I brought it up. I believe that Molly, like all the people in the group, has at least one very unsatisfied relationship in her life. In terms of this chapter, she's trying to relate to someone in her quality world and, because of external control, on her part or on the part of the other, she's frustrated.

For example, it may be her husband whom she pictures as treating her differently than she wants to be treated. Rather than accepting that she can't control him, she is trying to control herself and all around her "perfectly." She is trying to send the message, *I'm a perfect person; why don't you see that and treat me better?* But no one including she can be a perfect person and she has been given the warning of fibromyalgia to tell her this.

Since she can control her own behavior she has to change the perfect picture of herself that she has put into her quality world to a less perfect picture. She also has to change the picture she has of her husband in her quality world to a more realistic picture. If she can understand the concept of the quality world she can do this. Her quality world is her creation. She can change what she pictures in it, make it less demanding, and get along better with herself, her husband, and everyone else.

I realize this isn't easy but I also remember what a good friend of mine years ago said over and over: Consider the alternative. Molly may have to read this chapter over and over and she may have to get a little personal counseling, but the pictures she's inserted into her quality world have to be changed. And she's the only one who can change them. In the last chapter she made a start in this direction. The group has the example of Barry and Joan doing this same thing. As they got along better they changed the picture of each other and of Josh in their quality worlds.

Our quality world plays an important part in all our major decisions. It is called quality but each one of us determines what quality is at the time we put it in and we are not stuck with this initial standard unless we choose to be stuck with it. If we can understand how the quality world works, the odds are we will relax and allow this world to become more flexible.

If we keep it too rigid, as Molly seems to be doing, the pictures we keep can become more of a liability than an asset.

What is quality to us need not have anything to do with ordinary notions of what is right or moral. An alcoholic may put alcohol into his quality world; a business manager may put "creative" accounting into his. As I explained in Chapter 3, pleasurable pictures that have nothing to do with people may take precedence over our relationship pictures and, in the end, lead to a lot of unhappiness.

When we feel bad, we quickly turn to our quality world and see if we can satisfy one of its pictures in order to feel better. For example, an unhappy child doesn't have to grope around for what to do when he's unhappy. He immediately looks for someone he pictures as caring in his quality world. For most of us it's our mother followed closely by our father. After that, in no particular order, are our grandparents, brothers, or sisters. Unless we have very bad experiences with them, we almost always put relatives into our quality world where they are immediately available if we need them.

Later in life when we start to think about love and sex, we are on the lookout for men and women we picture in our quality world who might love us enough to put us into theirs. When that happens we have chosen to fall in love. But as I've already warned, that's the easy part. To stay in each other's quality world as sexual partners or family members and still not use external control is the hard part.

To find *things* to put into our quality world is easy. Acquiring them in the real world is what's so difficult. Early in life, we start to fill it with things we want, and to make sure we want them, there is a huge advertising industry continually pointing out to us how their products will help us fill that world. Companies who can figure out how to reach our quality world with their products make a lot of money. Advertisers don't neglect our basic needs. We are bombarded with products that will satisfy our need to survive, to love, and to gain power, freedom, and fun. The ad people may be more aware of our needs than we are.

Systems of belief are also widely promoted. Political beliefs, religious beliefs, moral beliefs, health beliefs, happiness beliefs, family beliefs, are all over the place. Choice theory is a system of belief that I am trying to teach you to use in your life if you want to be happy. But as I said before, no matter what you are trying to do with your life, it is the people in your quality world and your relationships with them that will determine whether you are happy or not.

It is especially important that we do what we can to persuade our children to keep us in their quality worlds and to put their teachers in them when they go to schools. Our homes and schools are bastions of the deadly habits that make this so difficult for our children to do. Our schools send notes home to parents blaming them for their children not behaving well in school. This blame compounds the problem; both the parents and the schools support each other in punishing students. Addiction, delinquency, and unwanted pregnancies are a natural outcome of our failure to persuade our children to put teachers and education into their quality worlds.

In order to be mentally healthy, we need at least one person in our quality world who is actively and positively involved in our lives. I've been fortunate. I've always had a group of people—family, friends, and colleagues—who are in my quality world. I also believe I'm in theirs. But short of explaining this concept and asking them, I'm not absolutely sure all of them have me in their quality worlds. But what makes me fairly sure is I don't practice external control with them.

Assuming you have had a similar assortment of important people in your quality world, how many of them do you believe have you in their quality world? To answer this question, let's look at Bev from the book club group. She talked about her daughter in a way that made it clear that this teenager was still very much in her quality world. Even though they were not getting along right then, there was no doubt she loved her daughter.

But from what Bev said, can she be sure she's still in her daughter's quality world? The words she used to describe her were, *This girl says she hates me and I'm beginning to believe her.* On the face of it, it would seem that Bev may not be in her daughter's quality world but my guess is she still is. It's very hard for a child to take a loving parent out of her quality world. But if Bev keeps using external control it could happen.

In the last chapter I said to Bev, *What I've been trying to teach you is that the problem between you and your daughter is you're both using external control. You're using it on her and she's using it on you. You both hate it. She's angering and you're depressing. The more you depress, the more she angers, because she believes the more upset and discouraged you are, the more she'll be able to get her own way. That's what she wants, isn't it? Her own way?*

When I said that, Bev gave me a look that seemed to say, *You really think I could get along better with my daughter?* Several members of the group said she should knock off the external control. What I am trying to explain here is that when we put a child into our quality world the attraction is we love them and we hope they love us. The only thing that can destroy that love is external control, but when it was suggested to Bev that she stop her use of it, she resisted, but less angrily than before. She thought that if she didn't try to control Brandi, she'd go down the drain.

Most parents, Roger and Jean excluded, have a picture of their relationship with their child that includes parental control. That control is an actual part of the parent's quality-world picture of how she should relate to her child. It's traditional, it's cultural, it's legal in many societies, it has the backing of most religions, and most of all, *it is the right thing to do.* We have not freed ourselves of this traditional quality-world picture in which, essentially, the parent owns the child. If a child starts doing something the parent doesn't like, tradition supports the idea that the parent should do all she can to change her back to the way she was.

The same thing is going on between Selma and Jim, her psychotic son. She is dedicating her life to trying to change him back to the child she pictured before he became psychotic. But when we use external control on anyone, especially when we are related to them, the people we use it on will resist. The closer the relationship the more they take exception to control and the more they will resist. It is likely that Jim's resistance is expressed through his psychotic symptoms. The *DSM-IV* is a huge book filled with resistant behaviors.

How You Picture Yourself in Your Own Quality World Is Crucial to Your Happiness

In my quality world, I have all my hopes, aspirations, and secret desires. I picture my wife, family, and colleagues respecting and supporting me, but I also picture myself achieving a great deal both professionally and socially. For example, when I entered medical school with the goal of becoming a psychiatrist, I pictured myself improving the mental health of the world. I have never given up on that picture. But until this book, I shared it only with my wife and a few very close colleagues. I was well aware that the world would frown on that much ambition in a young person.

In fact, whenever we set our aspirations for ourselves in our quality world too high, we push ourselves right into the jaws of external control. I did make a few small steps in this direction years ago, but nothing as drastic as I'm making in this book. Still, I invited a lot of criticism and had almost no support among my colleagues until I joined forces with Dr. Breggin and his group, the International Center for the Study of Psychiatry and Psychology, or ICSPP.[1] Now, with the support of this group, I am inspired to declare myself. But that does not mean the group has embraced choice theory. Their main thrust is to challenge the existence of mental illness as exemplified by *DSM-IV*. I do not, however, plan to choose any symptoms if my ideas are less accepted than I'm hoping for.

In the sixth grade, I didn't know anything about my quality world but I was aware that if I disclosed what I believed my relationship was with Miss Shehan, my classmates would have laughed and made fun of me. In that environment it would have been impossible for me to get everything I got from that year with her. I kept my mouth shut. In the external control world we live in, people satisfy their need for power by putting others down. Sharing your quality-world pictures makes you a target for put-downs.

Even before the sixth grade I had a fantasy life filled with ideas I can no longer remember. I'm sure many of you reading this book did, too. If someone had asked me about this life then, I wouldn't have been willing to disclose it. If, however, Miss Shehan had asked me, which she didn't, I might have considered telling her a little about it. When she did what she did, I knew I could trust her.

Keep in mind that you may not own anyone else, but you do own yourself in your own quality world. You can reduce your

[1]Peter and Ginger Breggin have stepped down from leading this group but will remain in close touch. Dominic Riccio is the new director. The address of the group is: 1036 Park Avenue, Suite 1B, New York, NY 10028.

expectations of other people and you can also reduce your expectations of yourself. It's not easy to do, but I know some women with fibromyalgia who have done it, and their pain has melted away. But they have to be vigilant. If they go back to aspiring to be perfect, the unhappiness and pain will return. It's also been my experience that some people with fibromyalgia, even when they learn what I'm explaining here, refuse to reduce their aspirations. They have decided that pain is preferable to accepting themselves as less than perfect.

For me, my quality world contains more than my hopes and aspirations. I see it as a place of refuge for my creativity and off-the-norm thinking. It is a place where I can enjoy the pleasure inherent in experiencing my own brain's possibilities without feeling I have to push myself to try to do more than I can do. But as I said, I'm still going to be cautious and careful with whom I share my outside-the-box thinking. Few of us are strong enough to deal with criticism and still be happy. It depletes our strength when we most need it.

I may fail in my effort to improve world mental health but I'm satisfied that I'm ready to reveal my ideas. If I didn't try, I'd be my own wet blanket and settle for less than I believe I have to offer. To get the full benefit from your quality world, you have to have the guts to open it up selectively. I don't advise that you ever wear your whole heart on your sleeve; a part of my heart I may never disclose. But if I keep these ideas all to myself, when I believe they're so needed, I undermine my confidence and I negate the very purpose of my quality world.

We Have to Be Careful Not to Overvalue Our Own Quality World

When people try to force their quality world on others, it leads to a lot of unhappiness. Proud, loving parents may have a picture of their child doing wonderful things with his life that he can't do or doesn't want to do and try to force that picture into their child's quality world.

For example, deciding on the basis of a little talent or good grades in science that the child needs to devote his life to music or go to medical school may be using the child to satisfy their own ambitions, and the child may recognize it and rebel. If this happens the intimacy of the child-parent relationship is lost. In some instances, the parent can be so devoted that the child may be afraid to express his own aspirations for fear the parent will ask him to take them too far. In either case, the parent-child relationship may be harmed.

I've heard major league baseball players on television say-ing they wouldn't expose their own children to the gung-ho, external-control coaching style rife on Little League teams. Without knowing anything about their child's quality world, they recognize he may lose interest in baseball if he has a bad experience with it as a child. Supporting the child's quality world rather than trying to change it is an important part of child rearing. If you see your child turn quiet and pay less attention to you, you may have gone too far in trying to con-trol your child's quality world.

The Quality World in Marriage

The best chance for sexual or marital happiness is for both partners to learn about the concept of the quality world as soon as they believe they are in love. This loving and accepting period is the time for each partner to put his or her quality world cards on the table. You may be in love but that doesn't make you naive. You know there are things in your quality world that will test your marriage. Disclose them and use your strong love for each other to work them out. Playing your cards from a closed hand before marriage or in marriage is very risky to your happiness.

As long as you have nothing to hide, it is easy to share your quality world with your mate or partner. But suppose one partner does something in the marriage that he knows the other won't accept. For example, the husband is unfaithful. What does a mentally healthy wife, who has a faithful husband picture in her quality world, do when she is faced with his indiscretion?

Here I'm treading on very sensitive ground. Nothing I can possibly say will please every reader of this book. The only way I can answer that question is to ask both partners to look back into their marriage and evaluate it in terms of external control.

For example, if neither the husband nor the wife is very controlling and you still don't get along well enough so that you are faithful to each other, I think you are incompatible and should consider divorce. For example, one may have had a much stronger need for freedom than the other and can't cope with the restrictions of marriage.

But relatively few marriages fail because of incompatibility. Almost all of them fail because of external control and the deadly habits. So, if after reading this far, the couple can see that their marriage is filled with criticism, blaming, complaining, and nagging and they still want to stay together, I'd advise them to try to get rid of the habits. Even though one infidelity has occurred, you may be able to save your marriage if you can both make an effort to give up the deadly habits. A little toleration is a small price to pay to get rid of the external control and save the marriage.

Sharing Your Quality World with Others besides a Mate

For example, you may have a picture in your quality world of your daughter remaining a virgin until she graduates from high school, certainly, a reasonable picture from your standpoint. You are preparing to use external control tactics such as grounding her to keep her away from a current boyfriend.

My suggestion is to take the time to explain the idea of the quality world to her and share some of your pictures with her. Tell her that you won't criticize her if she shares some of her quality-world pictures with you. As much as you can, it's important to let her know where you're coming from and that you know where she's at. Explain that it's hard for you to change; you've had your picture for a long time, not to restrict her, to protect her. But if she doesn't want to share her quality world, don't put any pressure on her to change her mind.

When you share some of your quality-world pictures, be sure to include some she might agree with, like a picture of her having a happy, successful life. She may then share some of her pictures with you and you may be able to see that there are positive things about her desire for love and sex. What you're trying to do is find enough common pictures in both your quality worlds so that you and she can keep close even though there is one major disagreement.

Instead of putting pressure on her to change her picture to yours, which she won't do, see if you can get an honest discussion going between you and attempt to negotiate the possibility that you'll consider changing some of your pictures so they are closer to hers. But here you have to be careful to avoid the seventh deadly habit, rewarding to control. It will be a real test of your mental health if you can do this. It may not work but in many instances there is nothing better. Only by doing something like this will your daughter elevate your place in her quality world. The more positively she pictures you, the better it will be for both of you.

This is what Roger and Jean did when Barry and Joan got involved with sex, alcohol, and drugs. They made their parental relationship with Joan and Barry take precedence over what they may have wanted. If you need a reminder of what Jean and Roger did, go back and reread Chapter 7. Joan, Barry, Roger, and Jean are all front and center in one another's quality world.

There used to be a program on television called *The Paper Chase*. In it a brilliant but insensitive professor, who had the picture of himself in his quality world as the world's greatest legal mind, bombarded his vulnerable students with criticism and sarcasm. It made for great TV—the core of all drama is external control—but I wouldn't like to have been a student in that class. What they learned from that experience may have led them to become effective lawyers but it did nothing toward helping them to become caring human beings. I cringed the few times while I was watching the program and I do all I can to teach in exactly the opposite way.

twelve

The Fifth Choice Theory Focus Group Session

Two more people joined the group but we have plenty of room around our dining room table. Of course, there was Joan, who'd been with her mother when we held the last session; and Neil, whose sister, Amy, has panic attacks. For the previous meeting he'd had to cancel at the last moment but it had been more for Amy than himself, he'd said when he called, and he'd be at this session. He said that Amy had told him she hadn't talked much about herself at the last session, so he wanted to be at this one to offer support.

Joan, who was coordinating all this, told me that several of them had gotten together after the meeting to rediscuss Chapter 7, about the session to which Roger had come with Joan and Barry, and Chapter 8, in which I'd introduced the ideas of basic needs, creativity, and our reasons for choosing symptoms like depressing. She wondered if I had any objection to their doing this. I told her to tell them the material was theirs to use in any way they wanted.

I did suggest that she tell them that if they brought in new people, they should have read all the material before they came to a session. She replied that they had already decided that and that a few members of the book club who hadn't read the material had asked for it. I said it was up to them, but if the group got too big they could split it. My idea was for Carleen and me to bow out after this meeting and to leave Joan and Barry nominally in charge. Roger said he'd help out if he was needed.

As with the prior meetings, I will remove all the small talk and kidding around that always takes place in groups that enjoy getting together. As soon as the initial small talk died down, Amy had a question: "I'm having a little trouble with your claim in the last chapter that to be happy we need to have at least one person in our quality world who plays an active part in our lives. I've been interacting for years with a person who's very much in my quality world but I'm far from happy. Am I missing something here?"

Her brother, Neil, who was sitting next to her, seemed puzzled when she said this, so I looked at him to see if he wanted to answer her question. He said, "Amy, you have a ton of people in your quality world who are active in your life. You fit that criterion perfectly."

Amy said, "Oh, I'm sorry. Sure I have. I guess all I was thinking about when I read that was Zack. He's my problem; everyone else is fine."

Bev said, "You hardly opened your mouth at our last meeting and now, all of a sudden, we hear about Zack. Who's Zack?"

Molly said, "If you'd tell us why you're so unhappy with him maybe we could help."

Jeff said, "Go ahead, Amy; you don't have to go into the gory details. We're your friends."

The whole group looked at Amy and she continued, "Zack is the man I've been living with for six years. We love each other. But I'm at the point where I want to get married and have a child and he doesn't. Even before we fell in love he told me that this is no world to bring a child into, but after we fell in love and moved in together, I never dreamt he still meant it. He says, if I want to get married, he'll marry me but he really doesn't want children. He's a kind man; he said he hates to see me unhappy and he's not an I-told-you-so kind of guy. He briefly reminded me that he'd explained this to me before we moved in together and that was all. But it's really hard for me to accept. I have my heart set on having a child. He's such a great uncle to our nieces and nephews, I know he'd be a good father. I'm miserable and I don't know what to do."

Neil said, "Amy, you've figured out what to do and you're doing it. That's why I wanted to come today. I knew you'd bring this up. I was surprised when you told me you didn't talk about Zack and your symptoms last time."

"You mean my panic attacks?"

Neil nodded in a very loving way and said, "You're unhappy, Amy; I'm your brother. I was afraid something like this was going to happen when you started to talk about getting married. You've been a little mother to every stray cat in the neighborhood since you were ten years old."

Roger looked at her and said, "Amy, I see a lot of young women with your problem. Do you remember what Bill said in Chapter Eleven about the difference between a sexual partner's quality world and a family member's quality world?"

"That chapter got me upset. I'll admit I didn't read it too well. All I could think of when I read it is if we're in each other's quality worlds, he should let me have a child."

Roger went on, "Bill said that the quality-world concept works differently for sexual partners than for family members. I had trouble with that idea and you're having trouble with it, too. Maybe Bill could explain it."

I said, "Remember what I wrote about family love being in our genes, but romantic love isn't? Sex is in both their genes and helps get people together but sex isn't love. It's part of our survival need. Amy's family love is genetic and it's driving her desire to have a child. That's why it's so strong. But Zack's love for her isn't genetic. It's romantic love and he doesn't want a baby. Actually, her love for him isn't in her genes, either. Their love for each other is romantic. It slips away as soon as external control enters the scene, and external control is taking over their relationship right now. This is why couples break up or get divorced. Families tend to stick it out. Bev will never 'divorce her daughter' and her daughter will never 'divorce her.' As long as they use external control they'll be miserable, but they'll stay together. Does anyone remember what I said about sexual partners and external control in Chapter Eleven on the quality world?"

Joan remembered, "I think it was something like for a sexual partner, we almost always change the criterion in our quality world to *I reserve the right to try to control you if I become dissatisfied with your behavior.* Unfortunately, reserving this right often kills the intimacy and ruins the love. That's what Barry and I were doing but we've cut way down. Amy, you're in trouble if you keep trying to make Zack do what he doesn't want to do. He'll take you out of his quality world and then you've lost him. My mom and dad never did this. Ask my dad; he's here."

Roger said, "I don't know exactly why, we just never did it. It's like Bill with Miss Shehan, we were lucky, we clicked. People call it chemistry but I don't think that's it. It's the total acceptance of the other; it's the intimacy Bill was talking about. But we weren't faced with Amy's problem. Intimate people do what Joan and Barry are doing now. They subordinate what each wants to the relationship. I have a woman like Amy in my practice now who doesn't want to do this."

I said, "Choice theory can explain a lot of things. It certainly explains this. Marriage is filled with quality world pictures that can't be reconciled. You want a child and Zack doesn't. You can't have half a child. There's no middle ground. But choice theory does offer another choice in this situation. It's a lot better than what she's choosing."

Amy immediately thought I was referring to her wanting a child and said, "I want a baby. What's wrong with that choice?"

Molly said, "No, not that, Amy. It's your panic attacks Bill's referring to. He's trying to tell you there's a better choice."

"Okay, okay, I'm panicking, you've all made that quite clear. Go ahead, tell me, what's a better choice? Maybe I can stop."

Joan said, "I don't know what a better choice is but I can tell you what I think would be a worse choice. I almost made it myself."

Selma said, "I know what Joan's talking about because I made that choice. I left my husband because of Jim. He was a good man and I turned against him because he refused to devote his life to Jim's mental illness like I have. So here I am still doing it and it hasn't helped me and it hasn't helped Jim to be without a father. Amy, from the way you're talking, you may break up with Zack over this. I'm not known for being that tolerant, but Zack has as much right not to want a child as you have to want one. You're miserable now. If you leave Zack over this, it's going to be worse."

When Selma said this, Amy burst into tears. Everyone got upset but Jill. She remained calm. After a few minutes, the group looked at her as if to inquire, Why aren't you upset for Amy? Barry actually said it, "Jill, don't you feel for Amy's situation?"

Jill said, "Truthfully, her tears don't impress me very much. Every day in my practice I'm faced with a lot more trouble than

Amy may ever have in her whole life. Dying of cancer when you're her age or losing a child to it is no picnic. I see it all the time. I agree with Selma. Amy has a good man but not a perfect man. I've never even had a good one much less a perfect one. I'm living with a woman now. It's not perfect, but it's better than with any man I've ever gotten involved with. Bill said you still have a choice. Why don't you stop feeling sorry for yourself and find out what it is?"

I said, "Amy, this city is filled with children who have no one to care for them. There are many ways you could get involved and mother them. They need you as much as the stray cats Neil mentioned. For all I know, if Zack's such a good uncle, he might be willing to adopt a child who's already here and needs a mother and a father. He might not go that far but he certainly wouldn't stop you from becoming a Big Sister to a teenage girl who may kill herself if she doesn't find some love. Besides, if you've got so much love, give it to someone you know who really needs it."

Amy had a blank look on her face, almost as if to say she didn't know what I was talking about. Jeff said, very gently, "Amy, Bill's talking about Zack. He needs you and you don't see it. The child in your quality world is crowding Zack out. I know what I'm talking about because I've been doing the same thing with women. All the women I pictured in my quality world were healthy, no arthritis. I picture myself the same way. But when a woman sees my hands she turns off. I've been through it over and over again. And you know what I've been doing? I've been turning against women just like I'm afraid you're turning against Zack. I'll never get this disease under control if I can't get into a good relationship, so recently I changed my picture. Three weeks ago I met a woman with arthritis and I'm falling for her. I think she's in the process of putting me into her quality world, arthritis and all. She's read all the material. She thinks Bill makes a lot of sense about most things. But I'll tell you, she thinks he's totally crazy where arthritis is concerned. I keep asking her to tell me what help have we gotten from all the arthritis experts that is so wonderful we should close our minds to another idea. I'd like to bring her to the group. Is it okay?"

Carleen said, "I'm sure it's okay with Bill and me."

The group all nodded in agreement. Jill said, "Jeff, I was thinking about the same thing. I have a young man about your age with arthritis and I was tempted to tell him about you and give him the material to read. I was thinking of inviting him to join us. I think you'd like him."

"Please do. Tell him about me and let him read the material. I'd like to get together with him and tell him about the group personally."

Barry said, "Bill, does it matter how big the group gets as long as they've all read the material and they want to come?"

Molly said, "But they've got to have read the material. I'm worried about what just happened here with Amy. Things got a little personal. It seemed like therapy to me and this isn't supposed to be therapy. I'm having a hard time accepting that I could be doing something to help my fibromyalgia. But I'm here because I haven't anything better. My doctor tells me I may be this way for the rest of my life. Then, in the last chapter, Bill said, I see myself as perfect in my quality world to avoid facing the fact that I'm not connecting the way I want to with an important person in my life. Amy is panicking. I'm going for perfection. Jeff wouldn't picture himself in love with an arthritic woman. That quality-world idea in the last chapter, I've been thinking a lot about that."

Neil said, "Molly, I was the one who told Amy she was choosing her panic attacks to deal with Zack, not Bill. Bill, would you have made that connection if no one else had?"

"No I wouldn't. Molly's right, that would be therapy. But you made it from the material you read. That kind of thing is exactly what I was hoping for. Use the choice theory to help yourself or help someone else. Connect with it yourself or try to connect it for someone else. That's your job, not mine. People learning to help themselves and each other is what I'm all about. Carleen and I are here to teach. We'd like you to use the ideas if they make sense to you. But if you do it or how you do it is up to you. I think what Molly just said was dead on."

Bev said, "But you did tell Amy about finding a way to satisfy her mothering need outside of having a child."

"You're right, but I consider that offering sensible advice, not therapy. I read Joyce Brothers in the paper and she often gives advice like that. Amy, did you object when I made that suggestion?"

Amy said, "No, that was fine, but I did get some counseling from Jeff when he said I should love Zack the way he is."

Jeff said, "It wasn't counseling; it was all right there in the last chapter. You love Zack. Bill said, if you've got a relationship, keeping it strong should take precedence over whatever else you want from each other. Look it up, Amy. Roger, you're helping a woman with that problem. Did you see that as therapy?"

Roger said, "I think, if we try to separate therapy from teaching something to each other, we're separating green apples from red apples. There are therapists who never give advice. I'm not one of them. Therapy has to include teaching and giving advice. What's the good of knowing all I know if I don't share it with my clients. But what it doesn't include is trying to force what you believe on your clients or anyone else. Amy would be wise to take Jeff's advice. My suggestion for anyone who thinks they've gotten good advice from a member of the group is to buy a box of candy for the whole group. Belgian chocolates are my favorite."

Carleen said, "I've been thinking. Except for Bev and

Selma, everyone here has been trying to get help for themselves. This book is also for parents who need help with their children. Selma, you have a difficult son. You've begun to use some of the material, you said so last time. What's happened?"

Selma said, "Last time when Jeff told me to relax and leave Jim alone, I was angry but I got over it and told him I appreciated his advice. Now I'm about ready to buy the group a box of candy. I'll tell you, though, when I first read the stuff about mental illness, I said to myself, Bill's all wrong. Jim is schizophrenic. He's got something wrong with his brain. But then I thought, What has that diagnosis done for Jim? And I began to have second thoughts. He was diagnosed when he was nineteen; now he's thirty. They told me he had a mental illness and the only thing that could help him are drugs. So first it was one drug, then another, and now he's on the fifth drug plus half a dozen others over the years to reduce the side effects of the one that's supposed to help him. With each drug they told me this is the one that'll work. But none of them works, and he hates them. They keep telling me to be patient, new drugs are coming on line. For years, the doctors haven't even talked to him. He can talk; he has some good days. All they keep telling me is that he'll never be normal again and I should make sure he takes his meds. Bill may not be right but what they're doing for him isn't right, either. At least what Bill's saying, he's unhappy and creative, makes some sense to me. The problem is, I'm afraid to take him off the drugs, and I don't know what else to do besides what I'm doing. But am I controlling him? Or is his craziness controlling me? As long as I saw him as mentally ill, those ideas never crossed my mind. Doing what Bill suggested, trying to get involved with him, not with his craziness, that's working. He used to get these sudden angry outbursts. He still has them, but not so often. I think it's because I'm just not paying attention to them."

Roger said, "Selma, I've struggled for years with parents like you and children like Jim. If it helps, you're not the only parent whose marriage has broken up in a situation like this. I was glad when Jeff told you to give him a break. I agree with Bill when he said that Jim needs to get involved in a place where the staff doesn't think he's mentally ill and doesn't talk to him as if he's mentally ill. There are places like that. I think you should look into them for Jim. Like Bill said, people do recover from the creative unhappiness that's called schizophrenia.[1]

Bev interrupted, "Are there places like that for my daughter, Brandi?"

Roger said, "I doubt it. But even if there are, I wouldn't consider any of them until you've stopped using external control on her at home. The places I've investigated for teenagers are filled with external control and psychiatric medications. What it said in the book about marriage goes for parents and kids. You never know how good your relationship could be until you make an effort to get rid of the external control."

Bev protested, "That's what everyone was suggesting the last time we met. But it's impossible. I tried, I really did. We had a few real good days, and then last weekend she left on Friday and didn't come home until two in the morning, Monday."

Amy looked Bev right in the eye and said in a very serious tone of voice, "Bev, did you hug her and kiss her and tell her you were glad to see her when she showed up? Maybe made her a cup of hot chocolate and talked with her as you tucked her into bed."

---

[1]Daniel B. Fisher, M.D., recovered from a diagnosis of schizophrenia, went to medical school, became a psychiatrist, and now runs the National Empowerment Center for people recovering and recovered from that diagnosis. Visit their Web site (power2u.org/what.html).

Bev raised her voice and said, "Hug and kiss her? Tuck her into bed? You've got to be kidding. I went berserk and screamed at her for two hours. We didn't come to serious blows, but I slapped her a few times with her pillow, and she grabbed it away from me and hit me back. How can you even suggest I hug and kiss a girl who treats me the way she does?"

Amy said, "I saw you nodding your head in agreement when they suggested I just keep loving Zack, a man who's lived with me and loved me for six years and now won't let me have a baby. But you know what, Bev, I'm going to do it. If he's ever going to agree to marry me and have a child, it'll be because I stop trying to control him. I've made up my mind about something Bill didn't mention. Zack is either in my quality world or he isn't. If he is, I love him with no reservations. If he isn't, I should leave him. Brandi is in your quality world, Bev, or you wouldn't be so upset. She needs your hugs and kisses desperately."

"But I've got to try to control her. If I don't, I don't know what she'll do."

Selma said, "My guess is she'll continue doing what she's been doing or worse. I controlled my husband right out the door. The more we talk, the more I think I'm still too controlling with Jim. I don't say anything but I keep looking at him the same way I did for years. My advice to you, Bev, is to stop asking her when she's coming home and hug and kiss her when she does. If I'd kissed Jim's father more, we'd still be married."

Bev said, "Dr. Glasser, you seem sane to me. Do you go along with what they're saying, hugging and kissing a girl like Brandi?"

"I've been going along with it for forty years, Bev. No reason to stop now. I love Amy's suggestion. If you start hugging and kissing Brandi when she comes in late, you better have a lot of Kleenex on hand to mop up all the tears, yours and hers."

Molly said, "You think we can all help ourselves? And Bev and Selma can help their kids." I said, "Of course I do. None of you are mentally ill. We all have choices. Joan and Barry didn't have to do what they did. But they did it and now they're happy. Bev and Selma can make better choices, too."

Before I could answer, Bev said, "But people don't behave the way you're suggesting. They don't. Suppose all of us change and it doesn't work. Then where will we be?"

I said, "That's a very good question. If this were therapy, I'd try to answer it. But it isn't therapy. It's information. In my personal experience it's helped a lot of people. I don't like answering a question with a question, but where will you be if you don't choose to use this book and don't choose to come to these sessions, Bev?"

"I don't know where I'll be but at least I'll have the satisfaction of knowing I tried to prevent her from destroying her life. It won't be my fault if she goes down the drain."

"I'll accept that answer. I don't see my job as trying to do any more than I've done, write what I've written, and meet with you to explain what you don't understand. I don't have to explain external control to anyone here. If Bev chooses to use it, I can say, I've tried to persuade her to embrace choice theory. That's all I can do."

Jeff said, "Wait a minute. Hold it. You've written a book; we've had two meetings. A lot has happened and you'll accept her answer. You're not saying we have to accept it, are you?"

"No control from me for Bev or anyone else. After tonight Carleen and I don't plan to be at every meeting. I'm certainly not going to start an argument with Bev. The rules of the study sessions are, if you don't want to accept what another member says, speak up, disagree, or do what you want. There's all the freedom of speech and choice you want here, just no therapy."

Jeff said, "Bev, you're willing to let Brandi go down the drain rather than hug and kiss her. I can't believe it. Are you serious?"

Bev said, "Are you willing to settle for your arthritis if what Bill suggests doesn't work? He's way out on a limb with that one. What does he know about rheumatoid arthritis?"

"I'll never settle for my arthritis. Bill's not forcing his ideas down my throat. If what he says doesn't help me, I won't blame him. But you have no right to settle for what you know doesn't work. My God, if you don't want to kiss her, send her over to Amy, I'll bet she'd be more than willing to give her a kiss and a hug."

Joan said, "No, I don't agree with this. Jeff, you're a good guy and you want to help but I—I don't know how to say this without using external control—but I don't think we should use any of it in these sessions. We can disagree with each other. We can suggest things to each other as you did with Amy. But to tell Bev she has no right to do what she's doing or not doing, that's external control. We shouldn't do that here. We all have the right to do what we want. The only person we know what's right for is ourselves."

Jeff said, "I'm sorry, Joan, you're right. I apologize, Bev. Isn't that what we're supposed to do when we slip back into external control? Will you accept my apology?"

"No, Jeff, you don't have to apologize. I'm the one who should apologize. I was trying to test Bill, to see if he'd slip back into external control. Instead I caught Jeff. You're getting to know me. See what my daughter has to put up with. It's a miracle she's not worse off than she is."

I said, "Anyone else want to answer Bev's question, where will you be if my ideas don't work?"

Neil said, "We'll be right where we were before we read the material. No matter what we do in here, we'll never be worse off unless we try to control each other. That's good enough for me."

The others nodded in agreement, so I said, "It's not that Carleen and I are going to disappear. We'll still be around to answer questions, but like I said, I don't think you need us anymore. I'll transcribe this session. And then when I write a few more chapters you'll have the whole book. I'd like to have a show of hands for how many of you want to continue meeting once a month. That interval seems about right."

All the hands went up. Bev said, "But we don't have a choice. We have to come back."

Jeff said, "Bev, why do you say that?"

"Because I've started to care about all of you. You and these ideas are part of my quality world. And you all know me well enough by now that if it's in my quality world, it's in everybody's. I think kissing and hugging my daughter when she straggles in after being gone for two days is crazy, really nuts. But I'm coming to the next session because I can feel that stupid idea creeping into my head. I think I'm losing control of my mind."

Jill said, "If you are, it's the best thing that's happened to you in a long time. Face it, Bev, your quality world is smarter than you are."

The group stopped and had a little meeting. They asked Barry and Joan to lead the group and Roger to be there when he could. Carleen and I were satisfied. After they left, Carleen and I had a chance to reflect on these two sessions. I said, "I'm going to miss that group. I think Bev said it for all of us when she said, 'You're all in my quality world.'"

Carleen said, "I think when people read the transcript of these two sessions they're going to say that these people seemed too sensible and too functional to have the kind of problems they claim to have."

"Why do you say that?"

"It's just that all the propaganda about mental illness

leads people to believe they can't do anything to help themselves. They're so dysfunctional they couldn't possibly read a book and then get into a group and learn what they've learned. Our critics will say they weren't crazy enough."

"I see what you mean, but my experience tells me that this was a very accurate portrait of what people who are wrongly labeled mentally ill can do for themselves. Of course, some of them may want to get additional personal counseling. That is why it would be wise for counseling professionals who agree with this book to offer to sponsor a group and do what Roger is willing to do if he has time, see some of the members privately. But if we are able to persuade unhappy people that they are not mentally ill and don't need brain drugs, there will be many times the work for psychologists, counselors, and social workers in private practice than there is now."

thirteen

Important Material from Al Siebert, Ph.D., and Anthony Black

Very few mental health professionals have gone through what Al Siebert experienced as a young man. Like myself, Al is a follower of Peter Breggin and is a member of the ICSPP. What follows, written by Al Siebert, is an all too common example of how psychiatry can be hazardous to your mental health. It is reproduced exactly as it was written for this book.

In March 1965, I had finished all my course work in the clinical psychology program at the University of Michigan and was starting my doctoral dissertation research. To support myself and my wife I was teaching part-time for the Psychology Department and working half-time as a staff psychologist at the Neuropsychiatric Institute (NPI) at the University of Michigan hospital.

At the urging of Alexander Giora, my supervisor at NPI, I applied for a two-year, postdoctoral National Institute of Mental Health (NIMH) fellowship at the Menninger Foundation in Topeka, Kansas. Martin Mayman, director of psychology training at Menninger, arranged for me to fly to Topeka for two days of interviews. My department chairman, Wilbert J. McKeachie, and others all rushed their letters of recommendation to Mayman. Several weeks later Mayman telephoned me and said I'd been selected to be one of the three NIMH Fellows to start their program in September.

The Menninger postdoctoral fellowship was a significant honor. My instructors and fellow graduate students were especially delighted because I was the first graduate in clinical psychology from Michigan to go to Menninger.

I completed my dissertation research quickly and passed my oral defense of it in June. With no more classes to teach, I arranged to work full-time at NPI for the summer. Now officially a "doctor," I was issued a long white coat with my name on the pocket. I enjoyed the new status. I liked feeling the difference in how people working at the hospital looked at me and listened to me. I was the same person as before, but their perception of me was different.

Looking ahead to the start of the Menninger program in September, I reread the description of the program philosophy. I found one passage especially attractive:

"The specific goals of the program are to foster a searching curiosity about clinical processes that will generate new research into the nature of these processes. The program allows Fellows a two-year 'moratorium' in which they are free to re-examine and reintegrate their theoretical, clinical, and professional skills."

I decided to get a head start. I started writing a list of questions:

Why am I now called a mental health professional when all my courses have been about mental illness? I've never had a class or even a lecture on mental health!

Why do psychiatrists put so many people into mental hospitals when there is so much evidence that hospitalization is not effective?

Why do psychiatrists always blame their patients when their recommended treatments don't work?

Why are there so many kinds of schizophrenia?

If there is no cure for schizophrenia, what don't psychiatrists and psychologists understand about schizophrenia?

I also wondered: Why do most people who want to make the world a better place focus on trying to get other people to change? Why don't they work on themselves?

With more personal time now available to me, I read *Atlas Shrugged*, by Ayn Rand. I was intrigued with her way of showing how people act in selfish ways while denying their selfish motives. Then an interesting coincidence occurred at NPI. A psychiatric resident complained to me about a patient. He said, "That patient refuses to believe I'm acting for her own good."

"Are you?" I asked.

"Of course."

"By working with her," I asked, "aren't you learning how to be a psychiatrist?"

"Yes."

"Won't you enjoy the prestige, money, and working conditions that psychiatrists have?"

"Yes."

"If you succeed with her, won't the nurses and supervisors think well of you?"

"Yes."

"If you help her, won't you gain her appreciation?"

"Probably."

"If you can get her out of the hospital, won't that help reduce your taxes?"

Nodding, he said, "Indirectly."

"And you want her to believe that you're working entirely for her own good?"

He clenched his jaw. "But I am!" he declared, then walked back into his office and yanked his door shut.

I became preoccupied wondering what internal mechanism blinds a person from recognizing their selfish motivation while insisting their efforts are unselfish. Sigmund Freud discovered some subconscious mechanisms that keep people from being fully conscious—mechanisms such as denial, repression, and projection. I wondered if there might be a hidden defense mechanism inside the character structure of people who force unwanted help on others. I decided to name the defense mechanism "charity" as an operational definition. "What is the motivation," I wondered, "that compels someone to force help on others despite protests that they don't want help?"

And I began to wonder, "What happens in the training of psychiatrists that creates this self-deception?"

A few days later, Frank, one of the psychiatric residents asked me to be present in his office when he met with one of his patients. I'd done the psychological testing of Tony, a twenty-one-year-old, unemployed auto factory worker. He'd gotten into a fistfight with his father, beaten him up, and driven off in his father's car. The father called the police, who located the car and arrested Tony for auto theft. The judge had sent Tony to the hospital for a psychiatric evaluation.

Tony's caseworker, Lois, was also present. Frank said sternly, "Tony, your behavior is sick. We can treat your problem, but you must accept that you're mentally ill before we can help you."

Tony shook his head. "I'm not a crazy person."

"You are mentally ill!" Frank said, "You must accept that."

"No, I'm not! You doctors are crazy if you think I'm nuts!"

"We've argued about this before. You must believe you're mentally ill or we can't help you."

Tony's face got red. His nostrils flared and he breathed faster. "I'm not mentally ill!"

"Yes, you are!" Frank said.

"No, I'm not!"

"Yes, you are!"

"No!"

Lois and I looked at each other. Frank gave up. Everyone sat in silence. Then Frank said, "The meeting is over." He nodded to the aide to take Tony back to the closed ward.

Frank sat dejectedly with his head down. Lois and I left without saying anything to him.

"I'm shocked by what I just witnessed," I said to Lois.

"Why?" she said. "The first thing a psychiatrist must do is convince the patient that he's mentally ill."

"That's what I'm shocked to learn. During graduate school, I read hundreds of articles and books on psychotherapy. None of them mention what we just saw! The research reports about psychotherapy are silent about a major variable! This means that most reports about psychiatric treatment are badly flawed!"

"Al, do you agree that Tony is mentally ill?"

"No." I said. "He's a kid with poor control over his impulses, but I've seen many like him. At the juvenile court where I worked, he would be typical. He's okay. His father used to beat him when he was a boy."

"He told me that."

"This is a case of another dumb parent not realizing that one day his son will be bigger and stronger and may decide to get even. No, Tony doesn't have a mental disorder."

Several days later we heard that Tony had escaped from the hospital. I stopped Frank in the hallway and asked him, "What happened the other day in your office? You seemed uptight. What's going on?"

Frank took a deep breath and shook his head. "Al, I'm trying to do what my supervisor tells me, but I don't like it. When I ask questions about why I should talk patients into believing they're mentally ill, he tells me to work it out with my therapist."

"I heard that you've been warned to be more cooperative."

"This is confidential."

"Sure."

"I met with my supervisor before that session with Tony. He told me that my 'case' was discussed at a senior staff meeting. He said if I didn't work out my problems and my resistance to the program, they would drop me."

"That's a heavy-duty threat."

"I tried to do what my supervisor told me to do with Tony, but I hated it. I'm worried. If I get dropped from here, it would be difficult to find another residency. I may have to give up psychiatry altogether. That's a lot of years wasted. I've taken out loans . . ." He paused for a while, then said, "Thank you for your concern," and walked away.

It shocked me to see that psychiatric residents who question what they are told to say and think risk being screened out of the profession. I saw that the training program for psychiatrists used a reduction of cognitive dissonance technique. When people cooperate in stating and defending a false belief, a certain percentage of them will gradually accept the belief as true.

I looked for opportunities to learn more.

A twenty-five-year-old man was admitted to NPI with the diagnosis "acute paranoid state." I arranged to be the psychologist who tested him.

In my office I asked him, "Why are you here in the hospital?"

He clenched his jaw. "My wife and family say I don't think right. They say I'm talking crazy. They pressured me into this place."

"You're a voluntary admission, aren't you?"

"Yes. It won't do any good, though. They're the ones who need a psychiatrist."

"Why do you say that?"

"I work in sales in a big company. Everyone is out for themselves. I don't like it. I don't like to pressure people or trick them into buying to put bucks in my pocket. The others seem to go for it. Selfish, clawing to get ahead. My boss says I have the wrong attitude. He rides me all the time."

"What's the problem with your family?"

"I talked about quitting and going to veterinarian school. I like animals. I'd like that work. My wife says I'm not thinking right. She wants me to stay in business and work up into management. She went to my parents and got them on her side."

"I still don't see the reason for your being here."

"They're upset because I started yelling at them about how selfish they are. My wife wants a husband who earns big money, owns a fancy home, and drives an expensive car. She doesn't want to be the wife of a veterinarian. They can't see how selfish they are by trying to make me fit into a slot so they can be happy. Everyone is telling me what I should think and what should make me happy."

"So you told them how selfish they are?"

"Yes. They couldn't take it. They insist they're only interested in my welfare." He leaned over and held his face in his hands.

This seemed to confirm the blind selfishness behind the compulsion to force "charity" on others, but where did the "paranoid" diagnosis come from?

I asked him, "Did you tell the admitting physician about them trying to make you think right?"

"Yes. Everyone's trying to brainwash me. My wife, parents, the sales manager. Everyone's trying to push their thinking into my head."

There it was! The psychiatric resident heard him saying, "People are trying to force thoughts into my mind." The resident had been trained to diagnose this as a symptom of paranoia instead of seeing the truth in the statement.

I asked, "What has your doctor said to you?"

"He doesn't listen. He says I must believe I'm mentally ill. It's crazy."

I was beginning to see that. Psychiatrists put a patient into a distressing bind when they say to a patient, "You must accept into your mind the thought that you are mentally ill because you believe people are trying to force thoughts into your mind."

I went for long walks by myself to think about what I was discovering. I was learning much about the training of psychiatrists and the actual practice of psychiatry that had not been covered in my courses. No psychotherapy research reports address whether or not patients were told they must believe that they are mentally ill. No research reports identify what percentage of patients forced to submit to therapy have been told it is for their own good. I was learning that a psychiatrist in training who is strongly motivated to acquire the status, income, and power of a doctor of psychiatry must set aside his or her critical thinking skills.

My wife was concerned about my preoccupations. I reassured her that I just needed time to sort out some professional dilemmas.

I wondered, "Where does self-deceptive unselfishness, this compulsion to force help on others, come from?" I speculated that people compelled to force unwanted help onto others are indirectly trying to build up feelings of esteem. They gain esteem from others by doing what "good" people do; they help people seen as sick, impoverished, or abnormal in some diminished way. Is it possible that the need to be surrounded by poorly esteemed people is why psychiatrists see mental illness all around them?

I wondered, "If someone declares that they are Jesus or the Virgin Mary, why do psychiatrists and others try so hard to remove that thought from the person's mind? Why do they lock the person up for years and force medications on them while claiming they are doing it for the person's own good? Why are psychiatrists compelled to stop other humans from enjoying extreme feelings of self-esteem? Is the perception of mental illness in another person a stress reaction in the observer?" I wondered how I could test my hypothesis.

An Experimental Interview

A fortunate coincidence occurred a few days later at NPI. At the morning report we heard that an eighteen-year-old woman had been brought in by her parents during the weekend. The psychiatric resident in charge of the patient said, "The parents told us that Molly claims God talked to her. My provisional diagnosis is that she is a paranoid schizophrenic. She is very withdrawn. She won't talk to me or the nurses."

Each morning we heard that Molly would not talk to anyone. She refused to go to recreational therapy, occupational therapy, or any ward activities. She stopped talking to her doctor. After two weeks of such reports, the senior supervising psychiatrist said to the resident, "This patient is not responding to our treatment milieu. She is so severely withdrawn, you should start the paperwork to have her committed to Ypsilanti State Hospital."

Most of the staff nodded in agreement. One of the staff psychiatrists said, "Molly is so severely paranoid schizophrenic, she will probably spend the rest of her life in the state hospital."

I saw that this could be an opportunity to test out my hypothesis. Since Molly was headed for a lifetime in the back ward of a state mental hospital, I didn't think I could do any harm. I obtained permission from the resident to do a few psychological tests on Molly before she was transferred. Then I arranged with the head nurse on the locked ward to see Molly in the ward dining room the next morning.

That night I mentally prepared myself for the next day. Professor Jim McConnell was my adviser my first two years in graduate school. He constantly challenged me to look at behaviors and their consequences. With this in mind, I developed four questions to use as guidelines:

What would happen if I just listen to her and don't allow my mind to put any psychiatric labels on her?

What would happen if I talk to her believing that she could turn out to be my best friend?

What would happen if I accept everything she reports about herself as being the truth?

What would happen if I question her to find out if there's a link between her self-esteem, the workings of her mind, and the way that others have been treating her?

During my interview with Molly the next day, she told me about the events that led up to her being in the hospital. She was an only child. She wanted her parents' love, but they didn't give her much. Just enough to give her hope she could get more.

She would come home from high school and volunteer to help with the housework, the cooking, and the dishes. But her mother rarely showed any appreciation. Her father had been a musician, so she took up the clarinet in high school. She thought this would please him. Her senior year she was chosen to be first chair in the high school orchestra.

"I was excited," she said. "I believed my father would be proud of me. When he came home that night I told him. But he got angry. He picked up my clarinet and smashed it across the kitchen table. He yelled at me, 'You'll never amount to anything.'"

"How did you feel after that?"

"Awful. I cried and cried. I knew my parents didn't love me."

"What happened after you graduated from high school?"

"I spent the summer with my boyfriend. At the end of summer I went to nursing school and he went to a university in a different city."

"Why did you chose nursing?"

"I thought the patients would like me for all the nice things I would do for them."

"What was nursing school like for you?"

"I kept to myself. I didn't make friends with other student nurses except for one. We had to study a lot. The third term I got my first clinical assignment. I was really looking forward to it." Molly looked down.

"What happened?" I asked.

"The two women in my room criticized me." Molly's face twisted in pain. "I couldn't do anything right for them."

"How did you feel when that happened?"

"Like the world was falling in. It was horrible." She dropped her head. "I ran away from school. I took a bus to where my boyfriend was at college. He came and met me at the bus station. We went to a coffee shop to talk. I said I wanted to come and live with him, but he said he wanted to date other girls. He said we could still be friends, but I should go home and write to him."

"How did you feel after that?"

"Awful lonely."

"What did you do?"

"I left school and went back home."

"How did your parents feel about that?"

"They didn't want me there."

"And you felt . . ."

"Lonely. I stayed in my room most the time."

"So your dad and mom didn't love you. The patients were critical. They didn't like you. Your boyfriend just wanted to be friends. Your parents didn't want you to come back to live with them."

Nodding. "Yes. There didn't seem to be anyone in the whole world who cared for me at all."

"What an awful feeling. . . . And then God spoke to you?"

"Yes," she said in a soft voice.

"How did you feel after God gave you the good news?"

Molly looked at me with a warm inner radiance and smiled. "I felt like the most special person in the whole world."

"That's a nice feeling, isn't it?"

"Yes, it is."

There it was. Confirmation of my hypothesis. Her feelings of esteem had been driven to an extreme low; some inner psychological mechanism reversed it to an extreme high.

Two days later, I went up to the locked ward to pick up a patient scheduled for testing. When Molly saw me, she walked up to me and said, "I've been thinking about what we talked about. I've been wondering. Do you think I imagined God's voice to make myself feel better?"

I was amazed. I didn't intend to do therapy, but she seemed to see the connection. "Perhaps," I said, shrugging and smiling.

At the morning reports during the next week, we heard that Molly was talking to people, participating in ward activities, dressing better, and wearing makeup. The plan to commit her was postponed. The supervising psychiatrist said, "This may be a case of spontaneous remission. You can never predict when it will happen."

During my last week at NPI, I was pleased to hear Molly's doctor announce at rounds that Molly appeared to be fully recovered, had been transferred to the open ward, and would be discharged soon. It fascinated me that a patient diagnosed as extremely paranoid schizophrenic suddenly got better, and they viewed the recovery as happening all by itself. No one was curious or asked questions.

My wife was becoming increasingly upset with my preoccupied state. She pleaded with me to talk about what I was thinking. I knew from past experience that she wouldn't be able to understand. She had not gone to college after high school and was a devout Catholic. We had a warm, happy marriage and got along very well as long as we avoided talking about our beliefs.

I warned her that she would be upset, but she pleaded with me to talk with her. Against my better judgment I gave in. I told her my hypothesis about how people with weak self-esteem could free themselves from indirectly doing things to gain esteem from others. I told her how freeing it could be to think, As far as I am concerned, I am the most valuable person who will ever exist.

She looked at me with horror. She said, "No! Only Jesus can think that!" She buried her face in her hands and started crying.

I tried to calm her. I told her about my interview with Molly and how it confirmed what I was trying to understand. Instead of being reassured, however, she remained distressed. She took our car and went to her church to talk to a priest.

Stress Reactions in the Minds of Beholders

Martin Mayman left Menninger during the summer. Sydney Smith took his place as acting director. Shortly after my wife and I arrived in Topeka and rented an apartment, Smith telephoned me and asked me to meet with him. He showed me my assigned office and explained about secretarial and other services. On the way back to his office he asked if I felt I could handle the pressures of the program. I assured him that I could.

Several days later, Smith telephoned me again. He said that my wife was going to start into psychotherapy. He said he wanted me to go with her for an interview with a psychiatrist at the outpatient clinic. I agreed, but felt suspicious. Spouses aren't interviewed when someone starts into individual psychotherapy.

At the clinic the receptionist sent both of us upstairs to see Dr. Farrell, a clinic psychiatrist. After several minutes of superficial talk, he said, "Mrs. Siebert, will you please go down to the waiting room? I want to speak to Dr. Siebert privately."

This confirmed my suspicion. A spouse is never interviewed first before a partner starts therapy. After she left, Farrell said, "We would like some help from you. What do you think is upsetting your wife?"

I paused, knowing that I was at a choice point in my life. I asked myself, "Should I be honest with him or should I only tell him what I think he can handle?" I decided that if I had to be deceptive to keep the fellowship, then it wasn't worth having. I said, "It could be some of my ideas."

His eyebrows went up. He didn't expect this to be so easy. "Tell me about some of these ideas," he said.

I laid it all out for him. I described my speculations about why a suppressed need for esteem compels people to force unwanted help onto others and about a defense mechanism I called "charity" that Freud hadn't seen. I told him about my experiment with a statement of high self-esteem, how that had led to my interview with Molly, and her rapid recovery from so-called paranoid schizophrenia. I explained how the perception of mental illness in others is mostly a stress reaction in the mind of the beholder. I described the bind that psychiatrists put a patient in when they say, "You must accept into your mind the thought that you are mentally ill because you believe people are trying to force thoughts into your mind."

"I've tried to explain all this to my wife," I said, "but she can't handle it."

"She is disturbed and depressed," he said. He glanced at the clock on his desk. "We're going to have to stop now. I have other appointments to keep."

He picked up his phone and asked the receptionist to send my wife up. As she seated herself next to me, he said to her, "You were right, Mrs. Siebert, he is mentally ill."

"What!" I blurted out. I was halfway expecting it, but I wanted so much for him to understand my ideas, the words came as a shock.

"You're the one who is mentally ill, Dr. Siebert, not your wife."

"I am not mentally ill."

"Your thought processes are loose."

"I'm going through a developmental transformation. It's healthy. I just don't have things sorted out well enough to present them well."

"I'm not going to argue. You are mentally ill and need to be in a mental institution."

"I do not need to be in an institution! I will appear any place you name to defend the accuracy of my ideas. I have a legitimate doctoral thesis."

"You must go in immediately. You are quite sick."

I started to say, "Well, you are going to have to . . ." but stopped myself. I knew that a person who goes into a mental hospital voluntarily can get out much more easily than a person who is committed. I saw that they were so emotionally distressed by my thinking, they would commit me if I didn't handle them right.

"This is quite a shock," I said. "Let me think about it. Say, am I going to be charged for this time with you?"

Farrell's eyes brightened. "That's a very good question, Dr. Siebert, very good indeed. No. We'll just write this off as professional courtesy."

"Professional courtesy?"

"Yes," he said with a warm, benevolent smile.

Driving back to our apartment, my wife confessed that Smith called me because she had telephoned him. She said her priest in Ann Arbor told her I was mentally ill and advised her to contact psychologists and psychiatrists here at Menninger and ask them to get help for me.

The next morning Smith telephoned and told me to come to his office.

At two o'clock, I walked into his office and sat down.

"We have decided that to protect our program," he said, "we have to let you go. We cannot trust you with our patients. How do you feel about that?"

"I believe that you have the right and the responsibility to do what you think is best for your program. I'll go figure out another way to present my ideas to people."

He stared at me for a moment. "We don't want you to do that. You need to stay here and go into the hospital. You are quite mentally ill."

"My ideas make sense if you'd listen. Look, I've been trying to analyze why psychiatrists can't stand a person speaking metaphorically about having high self-esteem when they say they are Napoleon or Jesus or Cleopatra. Have you ever realized that when someone says people are trying to force thoughts into his mind, psychiatrists then try to force the thought into his mind that he is mentally ill? The key is to not listen to the words but to observe the behavior and its consequences. There's a defense mechanism around esteem that Freud overlooked. The way to break free is to think—"

Smith stopped me. He shook his head. "Dr. Farrell went into that with you yesterday, and he decided that these ideas are symptoms of illness."

"He listened to me for less than twenty minutes. What is sick about my ideas?"

"You are mentally ill, and we have a reputation for knowing what we are talking about."

"That may be, but please give me a logical explanation."

"Your ideas make me feel uncomfortable. You should be in a mental hospital."

"Wait a minute! My ideas make you feel uncomfortable so I am the one who needs to be in a hospital?!"

He laughed. "Well, you are mentally ill and in need of help."

"I don't need anyone's help. I know exactly what's going on and will take care of myself as I always have."

He leaned forward. Smiling, he said, "If you stay here and work on your illness, you can reapply to the program when you are well."

"I do not have mental illness. I can logically explain and support everything I have been saying. Why won't anyone make an effort to listen to my ideas?"

"We recognize mental illness when we see it. Dr. Farrell says your thought processes are loose."

"I'm doing what the Menninger program says I should do! Haven't you read the program philosophy?" I got a head start questioning clinical assumptions. "Don't you recognize your defensive behavior? When Farrell asked me what I was thinking I answered honestly. I didn't come to you, you came to me. And look at everyone's reaction. No one says they disagree with me or that I should go away. Everyone feels upset and decides the way to deal with their distress is to eliminate thoughts from my mind that they can't handle. They insist that I need to be locked up."

"You do need to be in an institution."

"What I need is to be left alone. My only request is for people to get off my back and stop trying to help me."

"But we don't want you to do anything that might ruin your career. You do seem to have some good skills."

"And you're only doing this for my own good?"

"Yes."

"No selfish reasons motivate you?"

"Assuredly not."

"I can see we've reached a dead end," I said. I stood up and walked out of his office.

I drove home feeling fascinated and frustrated. I had told both Farrell and Smith that the perception of mental illness is mostly a stress reaction in the mind of the beholder, but they couldn't see how their reactions validated my assertion. Their lack of insight was perplexing. They were like members of a cult, their minds controlled by a delusional belief system.

Saturday morning Dr. Farrell telephoned me at home. "You are very sick, Dr. Siebert," he said. "You need to go into a hospital. You're a military veteran and there is an excellent Veterans Administration psychiatric hospital here in Topeka. The director of the hospital is a friend of mine. He has authorized your immediate admission. If I drive over and pick you up, will you let me take you over? There will be no cost to you."

Another choice point. I wondered if I should take off or go along with them. Three months of mental and emotional turmoil had me feeling tired and wrung out. I was in a new city with no one who knew me well. I knew how easy it was for psychiatrists to have someone involuntarily committed. In 1965 more people were in psychiatric hospitals for so-called mental illnesses than for all other medical conditions combined. I decided that by going in voluntarily, it would be much easier to get out later. I said, "Okay."

A while later, as we were driving to the hospital, he said, "You've made a wise decision. You'll be better off now." He drove with an artificial smile frozen on his face. His face glistened with sweat. His hands gripped the wheel so tightly his knuckles were white.

I said, "This will be an interesting experience for me. Anytime I go into a new situation I learn a great deal from it. This should prove to be the same. I know I'll learn a lot."

He glanced at me with a look of pity on his face. He seemed to be thinking, "You poor deluded soul. So out of contact with reality, you are happy and optimistic about your plight."

"Do you know about Dr. Viktor Frankl?" I asked.

"Logotherapy is too superficial," he said. "Analytical, depth psychology leads to better insights."

"I'm disappointed," I said, "that we didn't have longer to talk. I didn't have a good chance to tell you all about my ideas."

"Would you like to tell me more?"

"Yes. No one has given me a full chance yet. Would you be willing to come over and talk with me?"

He turned and smiled at me. "Yes, I'd like that."

I didn't believe him. He was doing what psychiatrists commonly do. Lies and deceptions are justified if it will get a person locked up without force. I'd like to see an article in a psychiatric journal justifying psychiatrists' habit of lying to people that they think are mentally ill—while assuring people they can trust the doctor. It's like trying to open a savings account by writing a bad check.

My Accelerated Postdoctoral Education

The admitting psychiatrist gave me a quick physical examination. He took my wallet, keys, and watch and sealed them in an envelope for keeping in the patients' safe. When I asked him what the admitting diagnosis was, he showed me the admissions form. It read "acute paranoid schizophrenia."

I spent Saturday and Sunday on a well-run ward. Several patients introduced themselves and I went with a small group for a workout at the gym. Monday afternoon I was transferred to a different ward. When I saw that the ward held very deteriorated, heavily medicated, chronic patients, I laughed. The Menninger and VA psychiatrists were trying to make me accept that their perception of me was more accurate than my own.

After breakfast on Tuesday morning an aide yelled, "Medications! Time for big red!" Patients began to line up by the nurses' station. As each one stepped forward the nurse looked at her chart to locate his name. She would then select a small paper cup from her tray and hand it to the man.

An aide would ask, "Juice or water?" The man would state his choice and be handed a cup. After he put the medications from the small cup in his mouth, he would swallow them down with the juice or water. Then he would drop the cup in a wastebasket and move away. Then the next patient would step up.

As the last man in line walked away, the nurse looked at her chart. "Siebert," she said to the aide. He yelled, "Siebert!"

I got up and walked over, wondering why they called my name.

The nurse held out a small cup to me. "Here's your medication," she said.

"I haven't asked for any medication."

"It's on the chart," an aide said. "You have to take it. Doctor's orders."

"I haven't seen a doctor. I don't want any medication. Who is the doctor who ordered it?"

"Dr. Baum, the ward doctor," the nurse said.

"I've never seen Dr. Baum. He isn't supposed to prescribe medication for a patient he's never seen. What's the medication?"

"Thorazine."

"How can a doctor who has never seen me order me to take a medication I don't want?"

Two aides edged toward me, getting ready to grab me.

"If you don't want to take it this way," the nurse said, "there are other ways you can take your medication."

I saw that they would use force if necessary. Make me take shots, maybe put me in an isolation room. I saw that my chances of successful resistance were zero. I reached for the cup the nurse held out to me. The aides relaxed and stepped back.

"Juice or water?" one of them asked.

"Juice."

I put the red capsule in my mouth. I tilted my head back and swallowed the juice letting the fluid carry the capsule with it.

I sat in a nearby chair as the nurse and aides put their cart and charts away. I asked myself, Why is this happening?

I saw that they were communicating some powerful messages: "We are so much in control of you we can force you to take into your body whatever we decide. You are powerless against us." And, "Since only the most severely mentally ill patients are placed on a back ward and required to take antipsychotic medications, you must accept our statements that you are severely mentally ill."

The dosage of Thorazine they forced me to take was too strong a dosage. I soon felt very weak and lethargic. Soon after taking it, I fell asleep in a chair in the dayroom. I awakened later to find myself drooling, my lips and tongue swollen.

Wednesday afternoon an aide took me to Dr. Baum's ward office. Baum apologized for the delay in seeing me, but I didn't respond. I felt angry. I'd been in the hospital five days without being seen by a doctor before now. He conducted a routine intake interview. When I asked him to stop or at least reduce the medication, he said he wanted me to continue this dosage. He said, "It is good for you."

Friday morning I made a decision to take myself off the medication. I needed to keep my mind and body in peak condition. When it was my turn, I held the pill under my tongue while I swallowed the juice. A few minutes later I went to the bathroom and flushed the pill down the toilet while I urinated.

I had chances to escape during the next several weeks, but decided to stay in the hospital until they held my admission case conference. The way the Menninger people reacted to my breakthrough insights was so bizarre, I anticipated that in future years they might try to deny what they did. I knew that if I stayed in the hospital until my case conference was held, my hospitalization would be documented in the permanent medical records of the VA system.

Each time I met with Baum, while he was working up my case, I asked him to stop the medications. He always refused, saying the Thorazine was good for me. I was inwardly amazed and amused, because he didn't know and couldn't tell that I'd stopped swallowing the medication many days before. He was playing a doctor/patient game with me that was unconnected to any medical reality.

During the next four weeks I learned more about psychiatry from the "accelerated postdoctoral education" that the Menninger people arranged for me than I would have from two years in their formal program. I experienced firsthand how psychiatric labeling prevents the professional staff from experiencing "patients" as real people. Psychiatric patients are talked to and treated much differently than patients in medical wards.

My case conference was held four weeks after my admission. I escaped from the hospital the next day. I had been a voluntary admission so I knew they could not send the police after me. I spent the night with my wife in our apartment. The next morning I telephoned Baum and arranged to come back to the hospital to sign an "against medical advice" discharge in exchange for my wallet, keys, and watch. I knew the system. It wouldn't look good for the hospital director to have to report that a patient escaped. Baum tried to persuade me to stay, but I refused. In the final moments before we parted, we stood looking at each other, he seeing me as a psychotic man refusing help, me seeing him as a man who had sold his mind and soul to a deluded profession.

Postscript

My disillusioning, transformational experience was the best thing that ever happened to me. What I went through closely matches Maslow's description of self-actualizing peak experiences. I was in a state of high consciousness knowing that everything was happening exactly as it should. It felt joyous to feel my mind breaking free from what Buddhists call "consensus reality." Resistance was not futile. I had not been assimilated into either side of the illusory struggle perpetuated by the mental health industry.

I returned home to Portland, Oregon, and decided to use my professional skills to research the inner nature of people so mentally healthy they are made stronger by adversity (Siebert, 1996). The life I chose for myself has been very satisfying and rewarding.

When I run across my old VA hospital record, I have mixed feelings. I chuckle when I read the diagnosis "Schizophrenic reaction, paranoid type, acute," and "Discharged AMA." At the same time, I feel sorry for people diagnosed as schizophrenic who don't know how to survive "help" forced on them that is frequently more harmful than beneficial.

When I consulted with an attorney, I learned that no psychiatrist has ever had to pay malpractice damages for mistakenly diagnosing a person as mentally ill. U.S. laws allow psychiatrists to diagnose anyone as mentally ill for whatever reason they think up with no risk of being found liable for mistakes made or for harm caused.

It is a mixed experience for me to be a witness during the end of the dark ages of the human mind. On the negative side, psychiatrists and psychologists frequently misrepresent what is known about schizophrenia to the public, and the media parrots what psychiatrists say with no semblance of critical thinking. For example, many psychiatrists are declaring that schizophrenia is a brain disease like Alzheimer's, Parkinson's, and multiple sclerosis. They say this, even though decades of research has established that from 20 to 30 percent of the people who go through a so-called schizophrenic experience eventually recover from the condition and can do so with no medications (Siebert, 1999), as did John Forbes Nash Jr. (Nasar, 1998).

There is much evidence that something is seriously wrong with psychiatry and it is a profession that lacks insight into its dysfunctional behavior. With rare exceptions (Perry, 1974, 1999), psychiatrists cannot distinguish between someone having a transformational breakthrough and an emotional breakdown.

On the positive side, I see that the human race has started an exciting transformation to its next level of development. Some of the restraints holding back the transformation will be broken when psychology researchers begin to study the motives, cognitive processes, and personality dynamics of psychiatrists and others in the "mental illness" industry (Siebert, 2000). For the sake of thousands of people with so-called schizophrenia who have been told that they have an incurable brain disease and are forced to take neurologically harmful drugs (Breggin and Cohen, 1999), I hope it happens soon.

References

Breggin, Peter, and David Cohen, (1999). *Your Drug May Be Your Problem: Why to Stop Taking Psychiatric Medications.* Reading, Mass.: Perseus Books.

Festinger, Leon. (1957). *A Theory of Cognitive Dissonance.* Stanford, Calif.: Stanford University Press.

Nasar, Sylvia (1998). *A Beautiful Mind.* New York: Simon and Schuster.

Perry, John W. (1974). *The Far Side of Madness.* Englewood Cliffs, N.J.: Prentice-Hall.

Perry, John W. (1999). *Trials of the Visionary Mind.* New York: State University of New York Press.

Siebert, Al (1996). *The Survivor Personality.* New York: Berkley/Perigee Books.

Siebert, Al (1999). "Brain Disease Hypothesis for Schizophrenia Disconfirmed by All Evidence." *Journal of Ethical Human Sciences and Services,* 1(2): 179–189.

Siebert, Al (2000). "How Non-Diagnostic Listening Led to a Rapid 'Recovery' from Paranoid Schizophrenia: What Is Wrong With Psychiatry?" *Journal of Humanistic Psychology,* 40(1): 34–58.

Siebert, Al. *A Schizophrenia Breakthrough* (unpublished book manuscript).

The Following Article "Prescription for Scandal: Biological Psychiatry's Faustian Pact," by Anthony Black, Speaks for Itself

The last few decades have witnessed an explosion in the use of psychiatric medication. Indeed, the omnipresence of legal brain-altering drugs in our society is such that, nowadays, it is rare for us not to know someone who is on them if we are not already taking them ourselves.

Moreover, and contrary to popular perception, a marked increase in the practice of electroshock therapy is accompanying this legal drug explosion. Prior to 1960 this biological psychiatric arsenal was confined mostly within the walls of the major psychiatric institutions. Since then, the biological genie has escaped the confines of the mental institution and taken up residence amidst the population at large.

One of the reasons for this psychiatric colonization of the normal stems from the increasingly intimate association between the multibillion-dollar-a-year psychopharmaceutical industry and institutional psychiatry. The latter's psychiatric

[2]Reproduced with author's permission from *Z Magazine,* September 2001.

journals, conventions, and professional associations are all substantially underwritten by the former.

Another reason is the rapid growth in Western society of an overarching philosophy of biological reductionism. This notion posits that, in studying any higher organizational entity, the whole can be explained by the parts, the complex by the simple, the higher by the lower. If you are depressed, it is because you have a biochemical imbalance, rather than, perhaps, that your life has no meaning. If one goes to war it is because of individual "aggressive genes," rather than your being the pawn of complex sociopolitical forces over which you have no control.

The idea that fundamentally new ontological properties and laws emerge at higher levels of an organization, each level of which demands its own language and theory for its description and analysis, is given short shrift in the reductionist paradigm.

A third and perhaps more ominous reason for the dramatic rise in the fortunes of biological psychiatry is that its proponents have waged a propaganda war on its behalf that is riven with pseudoscientific claims and evidential suppression.

They continue to claim, for instance, against substantial research to the contrary, that shock therapy is harmless. Needless to say, no psychiatrists have ever volunteered to test this hypothesis themselves. In this they are probably wise, since the original animal research (of the 1940s and 1950s) demonstrating undeniable brain damage was damning in this regard, as has been much of the subsequent human clinical data. All of this evidence, however, as well as the vociferous condemnation by a legion of former patients, has done nothing to squelch the practice of this jealously guarded symbol of the psychiatric profession's medical and legal authority.

Particularly disturbing are the demographic trends for this controversial procedure. In Canada and the United States, well over 100,000 people are subjected to electroshock every year. Over two-thirds of these patients are women and almost half are elderly.

Still, while ECT is one of the heavy weapons of the modern biopsych arsenal, the more usual workaday armament is drug therapy. The first is targeted on a population of thousands. The second on millions.

Here again, proponents make a number of bold claims. Perhaps the most scandalous of these is that drug therapy is safe.

In 1980, 25 years after the introduction of neuroleptic (antipsychotic) medication, an American Psychiatric Association task force report finally, grudgingly confirmed what a number of previously neglected studies had attempted to call attention to, namely, that roughly 40 percent of chronic users of these drugs went on to develop tardive dyskinesia, a Parkinsonian-like movement disorder indicative of permanent brain damage. Subsequent studies amplified these fears by pointing the finger at other permanent brain disorders caused by the neuroleptics. These included tardive akithisia, a highly debilitating anxiety and hyperactive movement disorder. All told, the latest evidence supports rates of neuroleptic-induced brain damage exceeding an astounding 5 percent per year of usage.

That for clearly psychotic patients there may be a cost-benefit tradeoff to consider with respect to whether or not to take these medications (perhaps, as a minimal maintenance dosage) is rendered moot by the fact that few if any of the patients so prescribed are, or ever have been, told of the potential cost. Moreover, these drugs are routinely employed in institutional settings on clients that are patently not psychotic.

Given this sobering tale, it might have been expected that biological psychiatry would exercise the cautionary principle in its future endeavors. This was not the case. Instead, the next round in psychiatry's legal drug trafficking campaign was launched on an unsuspecting public with all the same hubris, euphoria, and woefully inadequate experimental investigation as the first.

So began the antidepressant revolution. Actually, the word "revolution" is slightly misleading here, for some of the antidepressants, like the tricyclics and the monoamine oxidase inhibitors, have been around for quite a while. Long enough, in fact, to garner a shadowy reputation. The tricyclics, like Tofranil and Elavil, are known to have numerous side effects, induce severe withdrawal symptoms, and be extremely lethal in overdose. The MAO inhibitors are so dangerous that the maintenance of a special diet is necessary to avoid life-threatening cardiovascular reactions.

The minor tranquilizers, like Valium, have also been around for decades and are probably the most widely prescribed psychiatric medication. Technically, they are considered apart from the antidepressants by virtue of their central nervous system action. Nevertheless, they too are associated with a host of side effects in addition to being both highly addictive and lethal in combination with other drugs.

The word revolution, then, should rightly be reserved for the latest generation of antidepressants, the so-called selective serotonin reuptake inhibitors (SSRIs) and their hybrid kin. These include such brand names as Prozac, Paxil, and Zoloft. What is revolutionary about them is less their mode of action than the extraordinary media fanfare and scientific claims accompanying them. Though this is not the first time that a class of drugs has been alleged to specifically target the presumed biological cause of a complex psychological function (i.e., depression), they are the first to benefit from the notion that they might enhance the normal human condition as well.

The credibility of both these claims rests on the theory, widely embraced by the general public, that depression involves a well-defined point source, or sources, in the brain upon which antidepressant drugs act like magic bullets surgically targeting the offending region(s). Such a theory, however, seems never to have been burdened with the facts, for the overwhelming weight of clinical and physical evidence suggests that the drugs act, not by targeting any hypothetical depression center, but by blunting affect and emotion generally. They act, in other words, nonspecifically to block emotional (limbic system) and higher cognitive (frontal lobe) connection. They don't target anything other than a generalized splitting of psychic functioning.

Indeed, there is a clear line of reasoning that the sine qua non of their action is precisely their toxicity. In this they are related to alcohol, the pleasantly delirious effects of which derive largely from its toxicity and that, likewise, doesn't cure or target any mental dysfunction at all. A more telling analogy is to be seen in the comparison with cocaine and amphetamine, both of whose effects rely, in part, on their inhibition of the reuptake of serotonin. Ironically, it was cocaine that was first hailed as a miracle drug and panacea for psychic ills by Sigmund Freud at the turn of the century. That was until he personally discovered its physically destructive and addictive qualities.

The analogy can be carried further. Both cocaine and amphetamine impact additionally on the dopamine and adrenergic neurotransmitter systems. So do the SSRIs. Moreover, the claim that these drugs work functionally and specifically is further belied by the fact that the serotonin system itself ramifies throughout the brain and spinal cord. Curiously, in light of the widespread concern about biochemical imbalances in the brain, the only known such imbalances (apart from a few hormonal conditions like Cushing's syndrome and Graves' disease) are those caused by the drugs themselves. Lack of appreciation of this fact leads routinely to travesties in assigning cause and effect. The inevitable rebound reactions that ensue upon cessation of medication are often interpreted in circular fashion, by doctor and patient alike, as confirming evidence of the previously hypothesized biological abnormality.

It must be stated at this point that none of the foregoing is meant to suggest that genes and biochemistry have nothing at all to do with moods and behavior. Nor is it meant to espouse a belief in some sort of metaphysical mind/body dualism. I take it that the psyche is obviously based in a physical substrate, and that constitutional factors clearly influence everything, from temperament to potential intellectual limits. But to see biological parameters as framing human potential is a far cry from believing that we have uncovered, or that there even exist, specific, localized chemical substances of complex emotional and psychological states. It is furthermore naïve to suppose that these drugs could ever act in a functionally specific (i.e., fine-tuned) way, given what we know of the neurophysiological complexity of even the most "primitive" of brain processes (like temperatures and water regulation, for instance).

Even more naïve, however, is to suppose that tampering, on a daily basis for perhaps years, even decades, on end, with an organ as delicate and complex as the brain, is not inherently dangerous. Certainly our experience with the neuroleptics suggests otherwise.

Equally worrying is that basic neurophysiological principles clearly argue for the potential for permanent changes in physiology when the brain's dynamic homeostasis is chronically altered or upset. A number of animal studies involving amplification of the serotonin system have already demonstrated a compensatory down regulation of serotonin receptivity resulting in the permanent loss of serotonin receptors.

Also worrying is a recent report in the British medical journal the *Lancet*, describing how a group of scientists in the United States had scanned human brains and found damage to serotonin neurons, caused, they believe, by the street drug Ecstasy. Studies with monkeys have reinforced these results. Ecstasy is thought to work, at least in part, by boosting the serotonin system.

Still, biological psychiatrists will argue, and most people believe, that the SSRIs have undergone a rigorous battery of independent tests, trials, and experimental protocols under the auspices of the American FDA to ensure their efficacy and safety. Nothing could be further from the truth.

First of all, the experimental studies for these drugs are constructed, financed, and supervised entirely by the drug companies. Their vaunted independence is a complete myth.

Second, the timeline of the trials is so ludicrously short as to fly in the face of the most elementary scientific reasoning. Prozac, for instance, was released onto the market with only six weeks of clinical trials. In essence, anyone now taking the drug for more than six weeks is involved in his/her own study into its long-term effects.

Third, the experimental protocol and statistical design of many of these studies are a complete scandal in their own right. In the case of Prozac, among other statistical shenanigans: data were pooled from different sources, then manipulated into shape; relevant clinical groups were eliminated from participation; additional confounding medications were administered simultaneous to the test drug; the dropout rate of roughly 50 percent and the reasons for were never factored into the final results: and, finally, the total number of subjects that actually finished a placebo-controlled study was a mere 286.

It is natural to ask at this point, why, given their potential danger, we haven't witnessed an epidemic of adverse reactions and brain damage related to these new-generation drugs.

As far as the latter effect is concerned, "witnessed" is the operative term. The serotonin neurotransmitter system, unlike the dopamine system upon which the neuroleptics principally act, is not linked directly to the body's motor system; therefore any damage that may occur is likely to be much less visible over the short and intermediate run. Moreover, any emotional scarring or loss that does take place is likely, again, to be interpreted as part of the original hypothesized "biological" disorder. That said, it must be noted that the SSRIs do, in fact, also affect the dopamine and adrenergic systems, and, like the neuroleptics, they can be expected to exert a malign, if peripheral, influence on these structures as well. Evidence to this effect has already been documented.

In terms of bad reactions, the case against the SSRIs is on much firmer clinical ground. Following its release in 1988, for instance, a flood of Prozac horror stories hit the media. A deluge of lawsuits quickly followed, whilst Eli Lilly, its manufacturer, embarked on a massive lobbying and propaganda campaign to protect its $1 billion-a-year (1993) Prozac market.

Among the many pathological effects that Prozac appeared to induce or exacerbate were paranoia, compulsion, depression, suicidal ideation, and violence. Numerous bizarre gratuitous murders and suicides were credited to its influence, and a number of august journals, including *Lancet* and the *British National Formulary*, came out with confirming warnings about "suicidal ideation" and "violent behavior." Interestingly, this symptom cluster is typical of amphetamine psychosis, and by now, the well-known results of protracted stimulant overdose. Like amphetamine, Prozac is functionally a stimulant.

Apart from safety, yet another claim routinely made by proponents of the biological psychiatric paradigm is that the long-term effectiveness of medication for neurotic disorders is superior to that of traditional psychotherapy. Once again, this is a claim with little or no clinical evidence to back it up.

Indeed, a number of comprehensive reviews over the past decade have come out decisively in favor of psychotherapy. Common sense would hardly dictate otherwise, for by suggesting to people that they are merely biologically defective mechanisms capable of handling their emotional/psychospiritual crises only with the aid of a technological crutch, many of the fundamentals and principles of psychological healing are completely undermined. Encouraging patients to give up on personal growth and understanding in favor of pills is, apart from being a philosophy of despair, a recipe for emotional disaster. Helplessness is substituted for mastery, dependency for autonomy, and an unexamined life takes the place of self-discovery.

Moreover, at precisely the time of greatest need, the patient-cum-psychic adventurer is delivered up to a zombie-like state devoid of both mental acuity, and the capacity for deep feeling, self-awareness, and self-empathy. That biological psychiatry could so blithely trample underfoot such granite pillars of therapeutic common sense is chilling.

Even more chilling is the fact that the biological paradigm has expanded well beyond the confines of the adult population. For though most medicated adult patients can be said to be nominally voluntary, medicated children can in no way be so considered. It is curious that, in an era deluged with an avalanche of new statistics detailing the pervasiveness of childhood poverty, neglect, and abuse, the psychiatric profession has chosen to ignore the obvious psychosocial causes of most childhood behavioral disorders and has opted, instead, to crusade for the wholesale drugging of this involuntary population on the basis of totally unsubstantiated theories of biological causation.

There is hardly a shred of experimental evidence to buttress such trendy childhood "disease" entities as minimal brain dysfunction, learning disorder, or attention deficit hyperactive disorder. No underlying local organic malformation, physiological malfunction, or chemical basis has ever been clearly demonstrated for these syndromes and no well-controlled clinical studies have ever unequivocally supported them either. This has not stopped the escalating prescription of such stimulants as Ritalin and Dexedrine despite a host of negative side effects, including tics, spasms, growth suppression, and chronically elevated heart rates and blood pressure.

Naturally, the same dangers, the same potential for permanent damage, apply with respect to these medications as they do to all the others, with the added complication that here, the potential for harm is compounded by virtue of the drugs' interaction with the developing brain.

Increasingly, Prozac is also being given to children despite their never having been part of the original experimental protocol. The license for such practice derives from the fact that, once the FDA has approved a drug, there are few restrictions on how or to whom a doctor can prescribe it. In line with this practice, the antidepressants in general have become a jack-of-all-trades medication prescribed for everything from insomnia to migraine headache.

In stark contrast to this massive, state-sanctioned drug-laundering operation is the harshly punitive "war" the state wages against illegal drugs. Though beyond the scope of the present discussion, this fascinating paradox points up the concluding need to briefly confront some of the broader social implications of the biological psychiatric paradigm.

As part of its general philosophical stance, the biological paradigm is a conceptual formation with an implicit, highly ideological portrayal of the nature of "human nature." In this sense it is aimed at us all, for at the heart of any political philosophy will be found a conception tendentiously tailored of what it means to be human, and it is just this conception that the reductionist psychiatric model seeks to address in a manner which is neither progressive nor in any way new. Indeed, it is politically and culturally reactionary.

Politically, the notion that the laws of human behavior and mental functioning should be phrased predominantly in terms of biological parameters ineluctably invokes the specter of social Darwinism. For if our behavior is thought to be strictly biologically determined then it is immutable, our fates inevitable, and the status quo merely reflects the "laws of nature." It is then but a short step to the rationalization of the manifest inequalities of societal wealth and privilege. A sort of updated version of the divine right of kings in pseudoscientific jargon.

Culturally, the notion that we should conceive ourselves primarily as biochemical mechanisms is not only dangerously dehumanizing and spiritually stunting; it leads inevitably to both a dismissive and escapist attitude toward many genuinely psychological and social problems.

In having suborned, in other words, a substantial proportion of the population into believing their behaviors are dictated principally by their genes and their biochemistry, biological psychiatry has not only set back the psychological paradigm 100 years, it has also fanned the flames of a simplistic, reductionist view of human nature and of human society.

Psychiatry may have festooned itself with self-congratulatory laurels vis-à-vis its increasingly "scientific" and "objective" orientation, but ironically, it has moved ever further away from the true meaning of those terms. Having jettisoned the language and level of analysis necessary for an appropriate dialogue with its clientele, it is no longer capable of seeing itself in any remotely objective way.

Possessed by the reductionist demon, psychiatry today remains blind to its own historical contingency, to its own social, cultural, economic, and political conditioning. Unable to see that it, too, has a case history, it remains insensible to its own, quite advanced pathology.*

> *Anthony Black is a freelance writer, concentrating on international issues. He has published in many major papers in the Toronto area. Reprinted by permission of the author.*

* The material in the Appendix cites a great deal of the research to support the claims made in this article.

fourteen

You Have Finished the Book, Now What?

What You Can Do by Yourself—the First Option

You now have the information you need to give up using external control and to start using choice theory. The essence of that move is to say to yourself, "Whenever I am involved with another person, if we are close I can choose to behave in ways that will continue to keep us close; if we are not close, I can choose to behave in ways that will bring us closer." More than anything else you do, following this counsel will guide you toward mental health and a happy life. This choice is completely under your control. You can make it, no matter what anyone else does.

Because you have been exposed to so much external control, I suggest that you keep this book close at hand. Any time you feel unhappy, go to the appropriate chapter and reread it. Moving from external control to choice theory is not going to happen overnight. As familiar as I am with this theory, I still have to keep reminding myself that I have a way to go before using it is automatic. My wife, Carleen, tells people she's a recovering external controlaholic.

Since choice theory is a new way to live your life, be patient with yourself. When you slip into a deadly habit, apologize. If the person you are apologizing to seems puzzled, explain a little choice theory to her. From my experience, she will be curious and appreciate what you are trying to do. It can be a learning experience for both of you and will help you to move closer together. From this point, it's a matter of practice. The more you use choice theory, the more it will become part of your life.

What You Can Do with Another Person— the Second Option

Many people find it enjoyable to share this book with a friend or family member and discuss each chapter as they read it. Don't forget to read the Appendix, as it answers many more questions about the hazards of psychiatry to your mental health than I was able to include in the book. Even though you have already read this book, when you reread it with another person you'll be surprised at how much more you'll learn. Even though Carleen and I have been immersed in choice theory for many years, as I wrote and she edited, we both continued to learn more about this theory.

Form a Choice Theory Focus Group—the Third Option

Using this book as a text, Joan and Barry were able to persuade the members of their book club to get involved in a Choice Theory Focus Group. But starting with *Choice Theory* in 1998, I have written seven additional books, all of which expand the use of choice theory and can be used by readers and focus groups to supplement this book. Also, I can see a person who is interested in joining a focus group coming to the group to observe a session or two before reading this book. I can't guarantee that all groups will allow an observer, but I can see most of them doing this and then going further and asking if someone in the group will walk the observer through the book, someone in the group would lend his. I believe we should work hard to make these groups easily available at no cost to anyone who wants to participate.

In the group in the book, Carleen and I helped the group get started. But after two sessions, we turned the group over to Joan and Barry. I suggest that this be the practice for all groups when they start. Try to find a professional who knows choice theory to help you get started and offer a few sessions. Remember this is teaching and learning, not therapy. The participants can always find the answers to questions that come up by referring to the book or to other books cited later in this chapter.

The ideal organizations to sponsor one or more of these Choice Theory Focus Groups are the local mental health associations that currently dot the map across this country. Since the strength of these groups is in providing each participant with a chance to discuss how he is applying choice theory to his life and to learn from others, I believe it would be best to keep the numbers small. The group in this book is a good model for size but the ideal number of participants should be determined through experience. If a group gets too large it can easily spin off into two groups.

I believe that anyone who has read the book and agreed with its contents should be welcome, whether they are there for their own mental health, to learn how to help someone else to mental health, or both. The only limitation should be to exclude people who are not able to sit in a group and sensibly participate.

For example, in the group described in this book, Selma was welcome to participate so she could learn to use choice theory to help Jim, her unhappy son with psychotic symptoms. But Jim, or anyone with symptoms that prevent him from sharing in the give-and-take I've described, should be checked out by someone in the group to determine if he or she is ready to participate. A good analogy would be the process by which John Nash, who had done the work qualifying him for consideration for a Nobel Prize, was checked out by the Nobel Committee to see if he would be able to participate sensibly in the Nobel award presentation ceremony.

Each group will have to decide this for itself. But it is important to keep in mind that these are not therapy groups and not a place to which to refer people whose symptoms impair their contact with reality, in the hopes that they will be "cured."

Starting a Choice Theory Focus Group

Here are some questions that will help get the first meeting of a Choice Theory Focus Group off to a good start:

1. What, in your opinion, is a Choice Theory Focus Group?
2. What do you, personally, hope to accomplish by attending?
3. What do you think ought to happen in a focus group?
4. What do you feel strongly should not happen in the group?
5. How are you attempting to use choice theory in your life right now?

Since I led the focus groups in the book, I was able to get the groups started. But for all other focus groups the questions above will help you begin. The leader will know these questions, but if they are displayed in the room where the group is being held it will help to keep them in mind and new arrivals will see them as soon as they join the group.

There are many people well trained in choice theory or

willing to learn choice theory who might be willing to help a Choice Theory Focus Group get started. There are thousands of professionals who have already been certified in choice theory through the training program of the William Glasser Institute. Many others, like Roger, who is helping Joan and Barry with the group in the book, have read my books and put my ideas to work in their practices. I have already asked around and found that professionals are anxious to read this book and, when I have explained the focus groups to them, express an interest in volunteering to help. When no one is making money from a worthwhile project, there are usually a lot of capable people willing to volunteer.

As I have already suggested, I can see Choice Theory Focus Groups being started or sponsored by mental health associations whose board members have read this book. I am hoping that the volunteers who now help the mental health associations direct people to psychiatrists would also be interested in getting involved more directly with mental health through sponsoring and leading these focus groups. If they needed training in choice theory to help them do this, the William Glasser Institute could easily provide instructors. Keep in mind these are not therapy groups. No professional credentials would be necessary to lead a focus group.

Also, after you read this book and you want to lead a group, you might want to approach a mental health association and volunteer. This is an interesting program and I believe that once it gets started there will be no shortage of volunteers to help keep it going. For meeting places I suggest that you do as AA does, only you won't need as big a room. In Los Angeles County, a few years ago when I checked, there were nineteen hundred AA meetings being held every week. Surely there are plenty of places where you could meet.

Besides mental health associations, an HMO would be an

ideal organization to start one or more Choice Theory Focus Groups. Half the people who come to their doctor with medical symptoms have nothing physically wrong with them; their symptoms are caused by unhappy relationships. A medical professional who is concerned about this problem and has read this book might want to volunteer to get a group going.

Schools and colleges are filled with unhappy students, teachers, and parents who can't afford help and get far less than what they need. These organizations can easily find someone skilled who can volunteer to lead for a while and then turn the group over to one or more of its members. If the participants are minors, an adult leader will have to be present.

The waiting lists of public mental health clinics and social service agencies can also be a huge source of participants. The majority of the people waiting are fully capable of reading this book and participating. They can attend and still stay on the waiting list for personal counseling. As stated in Chapter 1, they can participate even if they are on medication or presently seeing a counselor. As long as they read the book and want to improve their mental health, they will be welcome. This no-cost mental health opportunity can be made immediately available. As they get help for their own unhappiness, many of them will refer others to this no-waiting, no-cost opportunity.

Probation and parole officers can refer their charges to these groups. For most of the people they see, there is nothing free and immediately available. If the officer reads the book and believes in its value, his own case load can provide members for several groups. Skilled leaders will soon emerge from the groups and, by leading, they can both help themselves and others.

A large source of referrals to these groups can also come from the military. The service personnel and their dependents have many unhappiness issues, and this no-waiting-list opportunity to learn to help themselves can be beneficial. The people who staff military prisons are already receiving choice theory and reality therapy instruction at Lackland Air Force Base in San Antonio. If choice theory ideas work well in prison, how many of those people might not be in prison if these groups had been available when the officers became aware that their soldiers were in need of help. Millions more could be kept from county, state, and federal prisons if they have this kind of help when they first get into trouble. If you add their dependents, the numbers getting help could grow to several million.

Many churches have a community service orientation and could easily provide the small room needed for a focus group. Church members, who struggle with unhappiness, after reading this book might get a group together for themselves. The minister could also put a group together and offer it to the congregation and even, at times, open it to the greater community.

A Choice Theory Focus Group for mental fitness could be analogous to an exercise class getting together for physical fitness. I see them as a safe, supportive refuge where people who would like to be happier can meet and support each other. Once you are acquainted with the book and join a group, you can attend as long or as often as you wish.

As in a physical fitness program, where you can work out with people in different stages of fitness and feel perfectly comfortable, you can learn to put choice theory to work in your life at your own pace. It is my vision that, in a few years, anyone anywhere in the country will be within a short distance of an ongoing or newly starting Choice Theory Focus Group.

Recently, even before I sent this book to my publisher, I had a chance to meet with a group of professional counselors from Connecticut involved in trying to improve community mental health. I asked them to define mental health and, as I expected, practically all talked about recovering from mental illness. Only one said it was helping people to become happier.

I then explained the thrust of this book: to teach mental health as something completely separate from what is now called mental illness. They became so excited that, after reading the manuscript, they have already arranged for me to present it at a state mental health conference in Madison, Connecticut. Look for such mental health conferences on our Web site.

The William Glasser Institute's Web site (www.wglasser.com) is planning to provide a chat room where you can participate in a Choice Theory Focus Group on line. Your group can be scattered all over the country or all over the world. Time in the chat room will be reserved for people who have read the book and want to talk about it with others who have read it. Instructors from the Institute, including, occasionally, Carleen and myself, will be scheduled on the chat line to answer questions.

As I have stated over and over, these are not therapy groups. I recognize that there will be people like Bev's daughter Brandi, or Selma's son Jim, who might benefit from seeing a counselor, but it will not be up to the group to send anyone to a counselor. But if a counselor is leading or advising a group, he or she can offer private counseling. While there are all kinds of counseling, I recommend finding a counselor who has read *Choice Theory* and does not see you, or your family member, as helpless, mentally ill, or in need of brain drugs.

A Final Word

I think in this book I have defended my claim that establishment psychiatry is hazardous to your mental health. I also believe that there will be no problem putting the Choice Theory Focus Groups into practice, since no diagnosis is needed to join a group and no professional skills are needed to lead one. But what may happen if this mental health program gets under way is that there will be an increase in the number of people who seek counseling. To get reimbursement by a third party, the counselor will have to write down a *DSM-IV* diagnosis. Since the symptoms the counselor is treating will be accurate, I see no difficulty in her doing it. But if she does it, she should be prepared to explain what she is doing to her client. If the client has questions, he should be referred to this book.

I am, however, concerned that I have not made clear what I would do if I were confronted with an individual who behaves in ways that appear to family, friends, and even police, to be a threat to himself or others. In that situation, I strongly believe that the individual needs immediate placement in a treatment facility where, if necessary, he is put in a secure environment. Since this is a legal procedure, my concern is that the present law requires a *DSM-IV* diagnosis, in order to compel a person to be treated. While the law does not specify any particular treatment, it is almost certain that the treatment will be strong brain drugs, which, as I have already explained, may quiet the patient by diminishing his creativity but will not improve his mental health.

All I can hope for in such cases is that the family members, who are concerned about his mental health, will read this book and prevail upon the court to recommend, or the treating psychiatrist to consider, teaching the patient choice theory as all or part of the treatment. If the patient is in a secure facility, there is no risk in trying what I suggest in this book. Learning choice theory in a warm, supportive, drug-free environment cannot make the patient worse and may help him to recover his mental health. In all other medical treatment, if there is an alternative, patients or their families are given a choice. This book provides a safe, effective alternative to current practice.

Additional Resources

Since 1996, when I changed the name of the ideas that I have been working on since the 1960s to choice theory, I have written nine books, including this one, all designed to help you use choice theory in your life and work. The first book, *Choice Theory,* (1998), is the basic book. In it I introduce all the major ideas except for the mental health material in this book.

The other eight books are specific applications of these ideas that may help you at home and at work. The books are available from William Glasser Inc. at the Institute address in the Appendix. Except for *Every Student Can Succeed,* which is only available through William Glasser Inc., all are available at bookstores or Amazon.com. At this time all my books, except for the few noted below, are published by HarperCollins in New York. Following are brief descriptions of the books I've written since *Choice Theory.*

The Language of Choice Theory, 1998

In the external control world we live in we use a great deal of external control language when we interact with the people in our lives. This language is particularly harmful to relationships. Here, I cite about forty-five examples of how you might replace the language you use now with the choice theory language that improves and preserves relationships. The following examples, taken from the book, cover the four relationships that are most vulnerable to external control. In the examples, external control comes first, followed by the choice theory alternative in italics.

Parent to Child

Do your homework now. I don't care what it is. You better do it or no TV tonight.

Okay, I'm not going to argue with you. Let's look the homework over together to see if you understand it. And I'll be right here to help you if you get stuck.

Love and Marriage

You said you'd call last night and you didn't. You better have a damn good reason why.

It's so good to hear from you. What's happening? Any news about that contract you've been working on so hard.

Teachers to Students

The next time you won't wait for your turn in tetherball, you're benched for a week.

I know you are having trouble waiting for your turn. Let's sit down here and talk while the others are having their turn. You're really good, you know. How did you learn to play so well?

Manager to Employee

Sales are down. There's nothing wrong with our product, so it's something you're doing or not doing. I think you've had it too easy. You've lost your hustle. I'd advise all of you to get going and bring in some better figures this month. Any questions?

I guess we've had so many good years that I've gotten complacent. I used to do a lot of things that I don't do anymore. I don't want to be a pain but I want to get a little more active, make some calls with you, sit down and figure out some new strategies. If I'm off base tell me. Things could be better, maybe our product is not as competitive as it used to be. Let's not make this a big deal, I really couldn't care less whose fault it is. I just want to do my part to solve the problem.

* * *

The words that come out of your mouth are important. Once they are out, you can't put them back in. This book may help you to be more aware of the external control language you use.

Counseling with Choice Theory: The New Reality Therapy, 2000

Many of you have been to a counselor or know people who have been to a counselor. I believe in counseling and I think you would be interested in how I counsel using and teaching choice theory in the process. This book is an update of the 1965 book *Reality Therapy*. It was published in hardcover under the title *Reality Therapy in Action*.

What Is This Thing Called Love: The Essential Book for the Single Woman, 2000

Moving from love to marriage is no longer as smooth as it used to be. Too many couples live together before marriage but find it difficult to agree on making a commitment to marriage. This book, which follows a woman through that process and shows how she used choice theory as a guide, can be very enlightening for both women (and also men) in this situation. (Published by William Glasser, Inc.)

Getting Together and Staying Together: Solving the Mystery of Marriage, 2000

Many marriages are in jeopardy because the couple does not know how to deal with their incompatibilities. This book, which applies choice theory to finding a compatible mate or to solving incompatibility problems after marriage, takes choice theory deep into the marital relationship.

Every Student Can Succeed, 2001

Right now there are at least ten Glasser Quality Schools in which every student does succeed. Almost all of them are public schools. The choice theory ideas in this book can help any school that wants to improve learning, eliminate discipline problems, and be filled with happy students and teachers moving toward success. With new federal funding for improving student achievement based on the 2002 Education Act, "No Student Left Behind," almost all low-achieving schools can now afford the cost of this training. This book, is only available through William Glasser Inc. at the Institute address.

Fibromyalgia: Help from a Completely New Perspective, 2001

By learning how to deal more effectively with their own creativity, six million women and a million men could reduce or get rid of their suffering. This approach would be valuable for any chronic pain, for example, migraine headaches that have no organic cause. Apply these same ideas to what is called schizophrenia or many other *DSM-IV* diagnostic categories and similar results become possible. (Published by William Glasser, Inc.)

Unhappy Teenagers: A Way for Parents and Teachers to Reach Them, 2002

No relationship, except marriage, is damaged more by external control than parent/teen relationships. This is so sad because, of all the human beings on earth, adolescents who are treated without external control are the most loving. This book applies choice theory to the teen/parent and teen/teacher relationship. The reward for trying these ideas with your teenager can start minutes after you read the first two chapters.

See Two Sessions of a Choice Theory Focus Group on Videotape

When the book was finished, Carleen and I had the idea that we could find some actors, give them the manuscript, and ask them to read it and then simulate a Choice Theory Focus Group. There was no script, what they did was all improvised. The result was more than we expected. We could see what was written in the book come to life in front of our eyes. The videotape is available through William Glasser, Inc. For information on obtaining this video, log on to wglasser.com.

Let Us Hear from You at wginst@wglasser.com

Nothing delights an author more than hearing from people who are trying to put his or her ideas to work in their lives. Information about the William Glasser Institute and the work we do is in the Appendix and on our Web site. In the Appendix, I include a description of other books by both people I know personally and people I would like to know. Their books have given me the information I needed to write the second chapter of this book. If you read their books, I'm sure they would appreciate hearing from you. Add your voice to theirs and ours. The mental illness believers and brain drug manufacturers are powerful, numerous, and a hazard to your mental health. The quest for mental health needs you. The essence of what it is to be truly human is at stake.

Appendix

Throughout this book, I made many statements to support my claim that psychiatry can be hazardous to your mental health. As promised, in this appendix, I cite some strong evidence for these claims. Much of this evidence refutes the claim of the psychiatric establishment that mental illness, as described in the *DSM-IV,* actually exists and can be successfully and safely treated with a brain drug. There is also a great deal of evidence to show that all brain drugs used to treat these nonexistent mental illnesses act on the brain in ways that harm its normal functioning. Once a brain drug gets into your brain, you then have a real mental illness that in many instances cannot be distinguished from Parkinson's disease.

Further evidence will be cited to show that the manufacturers of brain drugs spend millions of dollars on public relations campaigns to support their belief in "mental illness" and to sell the brain drugs they make to "cure" it. This disinformation campaign has been so successful that it is hard to find a dissenting voice, public or private, anywhere in the world.

Finally, even though I hardly mentioned it in the book, there is substantial evidence that a great deal of harm is done to your brain by the still widespread use of electroconvulsive therapy (ECT). Everything cited in this Appendix is written by highly competent researchers and is supported by reams of factual data. I advise you to read some or all of the evidence cited here to protect your and your family's mental health from what is now offered to you as the truth.

Unfortunately, there is little that you, as an individual, can do to refute this psychiatric establishment and drug company propaganda. The best way to protect yourself or your family is to demand counseling or to organize a Choice Theory Focus Group and to cite the evidence in this appendix as the reason why you are doing this.

The story from Dr. Al Siebert, along with books and two journals, all of which I have read, cite overwhelming evidence to back up the title of this book. If you want more evidence, there are hundreds of references cited in these sources that you can track down on your own. Beneath each reference book that follows, I will give a brief explanation of why I cite this source. You need no explanation for the following:

1. Breggin, Peter, M.D. *Toxic Psychiatry.* New York: St. Martin's Press, 1991.

    Peter R. Breggin is the world's leader in the effort to make the public aware of the harm psychiatry can do to you and your loved ones. He is the founder of the International Center for the Study of Psychiatry and Psychology (ICSPP), a center that has joined the battle against what he calls the toxic effect of psychiatry on the whole world. As you read this clear, detailed book, you will see the harm being done to you and your family, directly or indirectly, by what I have explained over and over as you've read my book. While this book is over ten years old, every word in it is as accurate today as it was when it was written. All humanity owes a debt to Dr. Breggin. He was among the first to stand up for your mental health against those who in their ignorance and greed may do it harm. For details, log on to his Web site (www.breggin.com).

2. Breggin, Peter, M.D., and David Cohen, Ph.D. *Your Drug May Be Your Problem*. Reading, Mass.: Perseus Books, 1999.

This completely up-to-date book goes into every detail of how you or your loved ones can be harmed by psychiatric diagnoses and brain drugs. It gives you information about what you can do instead of taking them, as well as accurate and important information on how to get off them. If you are concerned about a drug you are taking or about a drug a loved one is taking, this is the book to read.

3. Glenmullen, Joseph, M.D. *Prozac Backlash*. New York: Simon and Schuster/Touchstone, 2001.

In Part I of this book, "The Dangers of Prozac-Type Antidepressants," Dr. Glenmullen describes, in more detail than I have ever read, his extensive personal experience with the dangers of this class of brain drugs. If you or a loved one is on these drugs, his is a warning you should heed. Dr. Glenmullen is not yet ready to refute the concept of mental illness but what he explains and describes has value to you whether he does or not. I liked his book because he comes across as a caring, sensitive person. If you are looking for a counseling psychiatrist in the Boston area, you would be wise to contact him.

4. Gosden, Richard, Ph.D. *Punishing the Patient: How Psychiatrists Misunderstand and Mistreat Schizophrenia*. Victoria, Australia: Scribe, 2001.

When I read this book, I thought back to 1948, when I was a psychological intern at the Cleveland State Hospital in Newburg, Ohio, a suburb of Cleveland. There, they had a back ward in which about seventy women diagnosed with schizophrenia were kept stark naked in a large heated room. Several times a day they were hosed down and the excrement was washed away. They were fed from a coffee can with no utensils. Some had been in that ward for over sixty years and most had had extensive electroconvulsive therapy (ECT). A few had been given over a thousand shocks. Gosden details other horrible treatments, many designed to "shock" people back to sanity. What he describes that pertains to my book is that brain drugs have now replaced every treatment except ECT. The point of his book, which is so important, is that the way schizophrenia is handled now has not changed in substance. Now, as then, the afflicted people are not only treated as if they can't help themselves but are often dealt with as if they are a danger to themselves and others. (I refuted this hundred-year-old belief in the Preface of this book.)

Gosden also refutes this by pointing out many examples of how to show that when they are treated humanely, they do not need to be locked up or to have brain drugs forced upon them. He agrees with me that these are lonely, unhappy people and cites a lot of research to back this up.

From my experience, I do not agree with his conclusion, that schizophrenia is a spiritual/mystical emergency. That may explain some of their behavior but not as much as he believes. But whether he is right or wrong on that one point, this is a valuable book to read if anyone close to you has been given this diagnosis.

5. Johnstone, Lucy. *Users and Abusers of Psychiatry,* 2nd ed. London and Philadelphia: Routledge, 2000.

This book is similar to Breggin's *Toxic Psychiatry* in that it goes into almost encyclopedic detail. But unlike *Toxic Psychiatry,* which was written for professionals as well as clients, this book is written for you and your family. Certainly she did not exclude professionals, but in this book, you can find yourself and what's the best course of action to take. Written with great compassion by someone who has been involved in mental health for many years.

6. Lynch, Terry, M.D. *Beyond Prozac: Healing Mental Health Suffering Without Drugs.* Dublin, Ireland: Merino Books, 2001.
   This is the book that gave me the impetus to write the book you have just read. I can't speak of it too highly. If ever a man puts a human face on mental suffering and offers an optimistic message, Dr. Lynch is that man. I've read it twice and would not be surprised if I opened it again. I am so pleased he agreed to write the Foreword that began this book.

7. Whitaker, Robert. *Mad in America: Bad Science, Bad Medicine, and the Enduring Mistreatment of the Mentally Ill.* Cambridge, Mass.: Perseus Publishing, 2002.
   Written by a very careful medical reporter, this 2002 book provides a clear picture of how the drug companies buy favorable brain research results from a number of establishment psychiatrists to back up their claims that neuroleptic drugs are effective for what is diagnosed as schizophrenia. Citing unimpeachable evidence, he shows that these drugs not only don't help, they lock you into your symptoms and make recovery difficult or impossible.
   If you are taking any psychiatric medications, you should buy this book and read it carefully. He also cites evidence to support the conclusion that in the psychiatric community there is no agreement on what schizophrenia actually is or whether it is even a mental illness. The drugs he cites in his book are prescribed much more in America than in Europe to cure an illness now being diagnosed in increasing numbers every day. But to make money from the drugs, it is necessary to diagnose the illness. If you have

been given this diagnosis, you should read this book. If you don't want to do that, then at least read the following 1998 quote from the book.

Whitaker quotes Nancy Andreason, the editor of the *American Journal of Psychiatry,* the psychiatric establishment's major publication, saying: "Someday in the twenty-first century, after the human genome and human brain have been mapped, someone may need to organize a reverse Marshall Plan so the Europeans can save American science by helping us figure out who really has schizophrenia or what schizophrenia really is."

Besides the books cited above, there is also a journal that is devoted to the topic of this book. It is called *Ethical Human Sciences and Services: An International Journal of Critical Inquiry.* I cite three important issues. Volume 3, Number 2 (Summer 2001); Volume 3, Number 3 (Fall/Winter 2001); and Volume 4, Number 1 (Spring 2002). The publisher is Springer Publishing Company, 536 Broadway, New York, NY 10012. Tel. 212-431-4370.

All three of these seventy-five-page volumes are specifically devoted to articles, comments, and book reviews explaining in detail the inaccuracy of biological psychiatry, the psychiatry espoused by the psychiatric establishment, and the harm it can do to your brain and your mental health.

There are many articles that refute the diagnosis of mental illness. There are enlightening articles on the psychiatrist/drug company alliance to promote the diagnosis of mental illness. Other articles describe the buying of research favorable to brain drugs and the huge public relations efforts by drug companies to sell the use of brain drugs to mental health practitioners such as counselors, psychologists, and social workers, who cannot prescribe them and directly to the general public.

There are articles describing the damage ECT and can do

your brain, articles criticizing the forced administration of brain drugs to adults diagnosed as mentally ill, and articles questioning the drugging of nonconforming young children with medications like Ritalin and Prozac that have never been tested in long-term studies for the damage they can do to the child's developing brain. Finally, there's an article showing the flaws in the research that indicates that schizophrenia has a genetic etiology. There is no hard evidence to support that claim. Each article in these three journals cites many supporting studies that you can follow up for more information.

Your own and your family's mental health is at stake. The popular media will provide you with little, if any, of the information in this Appendix. Read some of this material. It can give you the information you need to encourage you to get involved in your own mental health or in helping someone else to help themselves toward better mental health.

Carleen, I, and the staff of the William Glasser Institute are ready to help you. Contact us at:

The William Glasser Institute
22024 Lassen Street, Suite 118
Chatsworth, CA 91311
Phone: 818-700-8000
Fax: 818-700-0555
E-mail: wginst@wglasser.com
Web site: www.wglasser.com

History and Information about the William Glasser Institute

In 1967, I founded the Institute for Reality Therapy for the purpose of teaching that approach to counselors, educators, managers, and literally anyone who worked with people. Since its inception, I have greatly expanded my thinking with the addition of choice theory and have applied that theory to almost every aspect of reality therapy. I have also extended the use of choice theory into the schools, as exemplified by the Quality School program, and into managing for quality in all other areas in which people are managed. My ideas are being applied to an entire community in Corning, New York.

With all these expansions and applications, I have gone so far beyond reality therapy that, for accuracy, I was encouraged to change the name of the institute to the William Glasser Institute. In 1996 I made the change so that anyone who is interested in any of my ideas and their application could easily contact us. Over the years, as our teaching and training have expanded, satellite organizations have been set up in many countries around the world.

The Institute, under the leadership of Linda Harshman, coordinates and monitors all training and serves as an information clearinghouse. My latest thinking is often made available through audiotapes, videotapes, and publications. The *International Journal of Reality Therapy** is the research arm of the Institute and serves as a vehicle through which its members can publish their works on new ways of using and teaching reality therapy.

As mentioned, the basic purpose of the William Glasser Institute is to provide training for professionals who want to use my ideas in their work with others. There are five parts to

---

*To subscribe to the journal, order back issues (not articles), or obtain copies of the resource guide, please contact Dr. Larry Litwack, Editor, *The International Journal of Reality Therapy*, 650 Laurel Avenue, 402, Highland Park, IL 60035, Phone (847) 681-0290, and E-mail llitwack@aol.com.

this training, which takes a minimum of eighteen months to complete: Basic Intensive Week, Basic Practicum, Advanced Intensive Week, Advanced Practicum, and the Certification Week. All of the instruction is done in small groups, and by explanation, discussion, and demonstration. Upon successful completion of the process, the individual is awarded a certificate that states he or she is Reality Therapy Certified. The certificate is not a license to practice counseling or psychotherapy. These practices are governed by the appropriate licensing authorities in various legal jurisdictions in North America and in other countries.

The Institute employs user-friendly people trained in choice theory, so if you contact us, you can be sure of a courteous response. It is my vision to teach choice theory to the world. I invite you to join me in this effort.

Index